BASIC AND CLINICAL APPLICATIONS OF FLOW CYTOMETRY

Developments in Oncology

KLUWER ACADEMIC PUBLISHERS-DORDRECHT/BOSTON/LONDON

BASIC AND CLINICAL APPLICATIONS OF FLOW CYTOMETRY

Proceedings of the
24th Annual Detroit Cancer Symposium
Detroit, Michigan, USA - April 30, May 1 and 2, 1992

Edited by

Frederick A. Valeriote, Ph.D.
Alexander Nakeff, Ph.D.
Manuel Valdivieso, M.D.

Wayne State University
School of Medicine
Detroit, Michigan

Kluwer Academic Publishers
Boston/Dordrecht/London

Distributors for North America:
Kluwer Academic Publishers
101 Philip Drive
Assinippi Park
Norwell, Massachusetts 02061 USA

Distributors for all other countries:
Kluwer Academic Publishers Group
Distribution Centre
Post Office Box 322
3300 AH Dordrecht, THE NETHERLANDS

Library of Congress Cataloging-in-Publication Data

A C.I.P. Catalogue record for this book is available from
the Library of Congress.

Printed on acid-free paper.

Printed in the United States of America

CONTENTS

LIST OF PARTICIPANTS

Ayad Al-Katib, M.D.
Division of Hematology and
Oncology
Wayne State University
P.O. Box 02188
Detroit, MI 48201

Kenneth D. Bauer, Ph.D.
McGaw Medical Center
Wesley Pavilion, Room 540
250 E. Superior Street
Chicago, IL 60611

G.J. Brakenhoff, Ph.D.
Department of Molecular Cell
Biology
University of Amsterdam
Plantage Muidergracht 14
NL-1018 TV Amsterdam
The Netherlands

Nigel Carter, D.Phil.
Department of Pathology
University of Cambridge
Tennis Court Road
Cambridge CB1 1QP
United Kingdom

L. Scott Cram, Ph.D.
Life Sciences Division
LS-DO, MS M881
Los Alamos National Laboratory
Los Alamos, New Mexico 87545

John Crissman, M.D.
Department of Pathology
Harper Hospital
3990 John R
Detroit, MI 48201

Lynn G. Dressler, M.A.
Division of Hematology and
Oncology
University of North Carolina
at Chapel Hill
3009 Old Clinic Bldg., CB 7305
Chapel Hill, NC 27599

John Ensley, M.D.
Division of Hematology and
Oncology
Wayne State University
P.O. Box 02188
Detroit, MI 48201

Gerard I. Evan, Ph.D.
Imperial Cancer Research Fund
Dominion House
Bartholomew Close
London EC1A 7BE
United Kingdom

David Hedley, M.D., Ph.D.
Department of Medicine and
Pathology
Princess Margaret Hospital
500 Sherbourne Street
Toronto, Ontario
M4X 1K9 Canada

Charles Hitchcock, M.D., Ph.D.
Department of Pathology
The Ohio State University
Columbus, Ohio 43210
Formerly Director of Flow
Cytometry Laboratory
Department of Cellular
Pathology
Armed Forces Institute of
Pathology
Washington, D.C. 20306

Joseph Kaplan, M.D.
Children's Hospital of
Michigan
3901 Beaubien
Detroit, MI 48201

Awtar Krishan, Ph.D.
Michigan Cancer Foundation
110 E. Warren
Detroit, MI 48202

Peter M. Lansdorp, Ph.D.
Terry Fox Laboratory
B.C. Cancer Research Center
601 W. 10th Avenue
Vancouver, B.C.
V5Z 1L3 Canada

James F. Leary, Ph.D.
The University of Texas
Medical Branch at Galveston
Department of Internal
Medicine
Sealy & Smith Professional
Bldg., Suite 724
Galveston, TX 77555

Stephen Lerman, Ph.D.
Department of
Immunology/Microbiology
Wayne State University
550 E. Canfield
Detroit, MI 48201

Michael R. Loken, Ph.D.
Bio Logics Consulting
29 San Juan Court
Los Altos, CA 94022

Michael Long, Ph.D.
Department of Pathology
Children's Hospital of
Michigan
3901 Beaubien
Detroit, MI 48201

Alexander Nakeff, Ph.D.
Division of Hematology and
Oncology
Wayne State University
P.O. Box 02188
Detroit, MI 48201

Kenneth Pienta, M.D., Ph.D.
Division of Hematology and
Oncology
Wayne State University
P.O. Box 02188
Detroit, MI 48201

Stuart Ratner, Ph.D.
Michigan Cancer Foundation
110 E. Warren
Detroit, MI 48202

Mario Roederer, Ph.D.
B007 Beckman Center
Department of Genetics
Stanford University Medical
Center
Stanford, CA 94305

Melvin Schindler, Ph.D.
Michigan State University
Department of Biochemistry
East Lansing, MI 48824

Howard M. Shapiro, M.D.
283 Highland Avenue
West Newton, MA 02165

Paul J. Smith, Ph.D.
M.R.C. Clinical Oncology Unit
The Medical School
Hills Road
Cambridge, CB2 2QH
United Kingdom

Carleton C. Stewart, Ph.D.
Roswell Park Memorial
Institute
666 Elm Street
Buffalo, New York 14263

D. Lansing Taylor, Ph.D.
Department of Biological
Science
Carnegie Mellon University
4400 5th Avenue
Pittsburgh, PA 15213

Manuel Valdivieso, M.D.
Division of Hematology and
Oncology
Wayne State University
P.O. Box 02188
Detroit, MI 48201

Frederick Valeriote, Ph.D.
Division of Hematology and
Oncology
Wayne State University
P.O. Box 02188
Detroit, MI 48201

Daniel Visscher, M.D.
Department of Pathology
Harper Hospital
3990 John R
Detroit, MI 48201

James Watson, M.D., PH.D.
M.R.C. Clinical Oncology Unit
The Medical School, Hills Road
Cambridge CB2 2QH
United Kingdom

Mark Zalupski, M.D.
Division of Hematology and
Oncology
Wayne State University
P.O. Box 02188
Detroit, MI   48201

ACKNOWLEDGEMENTS

Generous support from the following donors were critical to the success of this Symposium:

- Division of Hematology and Oncology, Wayne State University
- Harper Hospital
- Children's Hospital of Michigan
- AMAC, Inc.
- Becton-Dickinson Immunocytometry Systems
- Boehringer Mannheim Corporation
- Bristol-Myers U.S. Pharmaceutical and Nutritional Group
- Coulter Corporation
- Dako Corporation
- Glaxo Pharmaceuticals
- Immunex Corporation
- Meridian Instruments, Inc.
- Ortho Diagnostic Systems, Inc.
- Research & Diagnostic Systems, Inc.
- Sandoz Pharmaceuticals Corporation
- The Upjohn Company

The contents of the book were prepared for camera-ready offset printing with typing, layout and editorial assistance by Claudia A. Valeriote.

1

CYTOMETRY 2000: LASER PROBING THE FUTURE

Alexander Nakeff

"You've come a long way, baby!" is an adage that characterizes the substantial advances made in the applications of flow and image cytometry since their inception, some few decades ago. Prior to the last ten years, much of the progress had been of a technical nature, one that was focused on improving instrumenta- tion. At the present time, and probably even more so in the near future, the majority of effort has been, and will be, directed to biomedical applications aimed at measuring physiologically-relevant cell functions in both normal and cancerous tissue. A great deal of dis- cussion has centered on how best to utilize these data to alter the course of effective medical treatment in both a prospective and interactive fashion. There are several important but difficult issues that need to be addressed in future applications.

One pertains to the ability to quantitate measure- ments in physiologically-relevant terms; this entails an effective correlation with independent measurements, such as spectroscopy and HPLC. A second problem has to do with the variable degree of inherent heterogeneity within solid tumors, both in terms of various tumor cell subsets with different proliferative fractions, gene and oncoprotein expressions and DNA contents, and normal resident and inflammatory cells. Although multiple biopsies throughout a relatively-large tumor have been used to address this problem, its eventual solution may be more intractable than first realized. Furthermore, the ability to dissociate a solid tumor biopsy for analysis (particularly by flow cytometry) while preserving its inherent heterogeneity remains a difficult problem that entails extensive controls and biological insight into the pattern of tumor growth and its regulation. Lastly, the inherent complementariness of flow and image cytometry needs to be optimized. For example, although cellular measurements by flow are essentially "black box". This technology is rapid, multiparameter and capable of high-speed cell sorting. Image, on the other hand, permits the analysis of cells within the context of the parent tissue structure by utilizing familiar morphometric parameters while retaining multi-parameter capability; albeit at a much slower analysis rate than that by flow. However, both approaches continue to present difficulties in signal quantitation. Nevertheless, the ability of analytical cytometry to measure multiple functional markers on single cells, simultaneously, rapidly and quantitatively, represents the untapped potential for exploitation of this technology in oncology. This includes the measurement of tumor growth, prior to, during and following different modalities of anti-cancer treatments, the presence of treatment-resistant sub-sets and the degree of micro-metastasis by rare-cell sorting and analysis.

The focus of this 24th Annual Cancer Symposium was on a substansive and free-ranging examination over a three-day period that was devoted to present applications of analytical cytometry to our understandings of the basic biology and clinical treatment of cancer, and speculation as to where we may be headed in the next millenium. Certainly, there will be substantial refinements and increased sophistication of flow instrumentation with more utilitarian software analysis and graphic packages; the emphasis may well be towards "smaller (custom, lower-powered lasers) is better" with analysis and sorting hardware being designed for more specific oncologic applications. New lasers (combining visible and ultra-violet spectra) and fluorescent dyes (e.g. in the far red) will continue to be developed, in addition to newer hardware and software approaches to high-speed cell sorting of rare-cell populations. In image cytometry, more sensitive cameras and screens will become available with improved software controls to permit high-speed analysis and possible quantitation. These advances, though impressive in scope, may well be dwarfed by the imaginative biomedical applications that we garnered a glimpse of at this meeting.

The proceedings were aptly opened and concluded by two pioneer "flow philosophers"; namely, Drs. Howard M. Shapiro and James V. Watson who have made major advances in the application of flow cytometry in basic and clinical oncology research, respectively. The excitement and images of the future raised by these two talented and far-sighted investigators, together with the body of innovative work summarized in these Proceedings, bear witness to the truism that:

"The creativity in the future applications of analytical cytometry is limited solely by the imagination of the investigator."

ACTIVATION AND PROLIFERATION OF PURIFIED HEMATOPOIETIC
PRECURSOR CELLS FROM HUMAN BONE MARROW

P. Lansdorp, W. Dragowska and T. Thomas

INTRODUCTION

Despite increasing knowledge of the phenotypic and
functional heterogeneity of early hematopoietic cells,
many questions regarding the biological properties of
primitive hematopoietic cells and their regulation
remain. Some of these questions are related to the
mechanism(s) by which hematopoiesis is maintained throughout the life-span of an individual. Is stem cell
self-renewal involved in this process or is there a
gradual depletion of an extensive stem cell reserve?
The answer to this question is likely to dramatically
impact on therapeutic options and strategies. For
example, if pluripotent hematopoietic stem cells would
have an intrinsic ability to self-renew (as is often
assumed) then in vitro conditions could likely be defined that would allow such cells to display this self-
renewal potential. This would in turn allow numerical
expansion of stem cells ex vivo for transplantation and
gene transfer experiments starting with very few (in
principle a single) stem cells. Such cells could be
obtained from relatively small amounts of either bone
marrow or peripheral blood and would probably be purified to homogeneity for the occasion, especially if such
normal cells were diluted by abnormal (i.e. leukemic)
stem cells. Suspensions of cultured normal stem cells
could then be used to ensure rapid engraftment, and such

cells would likely be the preferred source of cells in
settings where bone marrow transplantation is currently
used and gene therapy is being considered. If, on the
other hand, hematopoiesis is maintained by clonal suc-
cession of stem cells with a limited proliferation
potential, current efforts related to the harvest and
purification of stem cells in large numbers would not be
lost, as for safe and cost effective transplantation, a
relatively large number of primary cells would continue
to be desirable and required. In the hope of eventually
answering this important question regarding maintenance
of hematopoiesis, we have chosen to try to <u>directly</u>
study the biological properties of highly purified
hematopoietic progenitor cells ("stem cell candidates")
<u>in vitro</u> and <u>in vivo</u>. In this report we summarize our
strategy for the purification of stem cell "candidates"
from human bone marrow and present results of <u>in vitro</u>
studies with purified $CD34^+CD45RA^{lo}CD71^{lo}$ bone marrow
cells.

MATERIALS AND METHODS
Purification of Hematopoietic Cells

Suspensions of previously frozen bone marrow cells
retrieved from vertebral bodies of cadaveric organ
donors were used in these studies as a convenient and
reproducible source of hematopoietic cells (1). Cells
were washed once and resuspended in Hank's Hepes Buffered Salt Solution containing 2% FCS and 0.1% sodium
azide (HFN) for subsequent staining with monoclonal
antibodies specific for CD34 (8G12), CD45RA (8D2) and
CD71 (OKT9). These antibodies were labeled, respectively, with cyanine 5-succinimidyl ester (Cy5), R-Phycoery-
thrin (RPE) (2) and fluorescein isothiocynate. Cells
were washed twice in HFN and resuspended in HFN contain-
ing Propidium Iodide prior to sorting. $CD34^+CD45RA^{lo}CD-71^{lo}$ cells ("stem cell candidates") were sorted on a

FACStar$^+$ cell-sorter (Becton Dickinson, San Jose CA) and were collected in serum-free medium (see below).

Culture Conditions

The serum-free medium used in this study was a modification of serum-free medium described for murine erythroid progenitor cells (3) and was described previously (1). This medium is based on Iscove's modified Dulbecco's medium and contains 2% bovine serum albumin as well as insulin (10 μg/ml), transferrin (200 μg/ml), low density lipoproteins (40 μg/ml) and 5x10^{-5}M ß-mercapthoethanol. Cells were cultured in freshly prepared serum-free medium which, for some experiments, was supplemented with a combination of recombinant human growth factors including IL-6 (10 ng/ml), IL-3 (20 ng/ml), mast cell growth factor (MGF, a c-kit ligand, 50 ng/ml) and Erythropoietin (Epo, 2 U/ml) as previously described (1). After the indicated time intervals, the cells present in the cultures were counted, relabelled with antibodies specific for CD34 and CD71. CD34$^+$CD71^{lo} cells were re-sorted and used to initiate subcultures. This process was repeated up to five times after which too few CD34$^+$CD71^{lo} cells were recovered to continue the experiments.

RESULTS AND DISCUSSION

The strategy for purification of human "stem cell candidates" from bone marrow aspirates or cells extracted from the vertebral bodies of organ donors has been previously described (1) and is illustrated in Fig. 1. In those previous studies it was shown that CD34$^+$CD45RA^{-}^{lo} CD71^{lo} cells are highly enriched (at a good yield) in cells capable of initiating long-term cultures (as measured by the production of colony-forming cells after 5-8 wks on irradiated stromal cells (4) and that such cells are relatively depleted of myeloid (CFU-GM, CFU-G,

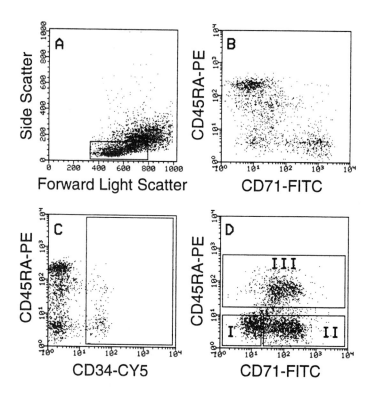

Figure 1. Sort windows for selection of stem cell "candidates" from human bone marrow. Low-density (e.g.
Ficolled) bone marrow cells from the indicated light
scatter and CD34 window (boxed areas in A and C) were
sorted with respect to CD45RA and CD71 expression as
shown in the boxed area in D (fraction I). The correlated expression of CD45RA and CD71 on all viable (PI-
negative) cells in the selected light scatter window (A)
is shown for comparison in B. From several vials of
frozen cells (10^8 cells/vial) a total of $20-100\times10^3$
$CD34^+CD45RA^{lo}CD71^{lo}$ stem cell "candidates" can be purified in a sort of several hours.

CFU-M, CFU-Eo) and erythroid (BFU-e) colony-forming
cells (1). The distribution of LTC-LC and clonogenic
cells among $CD34^+$ cells that were separated based on
CD45RA and CD71 expression is shown in Fig. 2.
Serum-free suspension cultures of $CD34^+CD45RA^{lo}$-
$CD71^{lo}$ cells in the presence of IL-6, IL-3, MGF, and Epo
for 10 days resulted in a large increase of cells, the

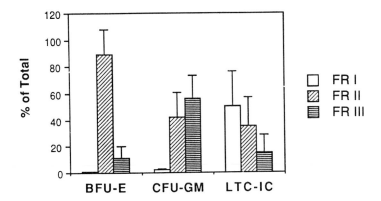

Figure 2. Differential expression of CD45RA and CD71 on functionally distinct progenitor cell populations. CD34[+] cells were sorted on the basis of CD45RA and CD71 expression as is shown in Fig. 1D and the purified cells were plated in various assays. Note that BFU-e are almost exclusively recovered in the CD34[+]CD45RA[lo]CD71[+] fraction, that the majority of myeloid colony-forming cells were found in the CD34[+]CD45RA[+]CD71[+] fraction and that the majority of the LTC-IC were recovered in the CD34[+]CD45RA[lo]CD71[lo] cell fraction.

majority of which were CD34[-] (Fig. 3). When CD34[+]CD71[lo] cells were sorted from expanded cultures (boxed area Fig. 3) and recultured, it was found that extensive cell production was observed again. Interestingly, the number of CD34[+]CD71[lo] cells in the (sub-)cultures corresponded very well with the number of CD34[+]45RA[lo]CD71[lo] cells that was used to initiate the cultures and the absolute number of CD34[+]CD71[lo] cells remained more or less constant throughout the culture period (Fig. 4). The observed balance between production and maintenance of cells could be indicative of a biological mechanism that limits expansion and yet avoids depletion of a primitive hematopoietic precursor cell pool. Current studies focus on the mechanism of CD34[+]CD71[lo] cell maintenance in these cultures using two different approaches. In the first, CD34[+]CD45RA[lo]CD71[lo] cells are labelled with the fluorescent tracer dye PKH26 (5) and PKH26

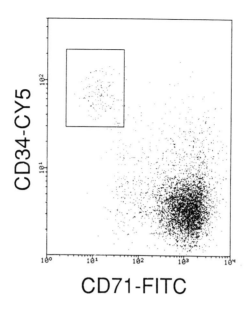

Figure 3. Cells with a CD34$^+$CD71^{lo} phenotype remain present in rapidly proliferating serum-free cultures containing IL-6, IL-3, MGF and Epo. Cultures were initiated with CD34$^+$CD45RA^{lo}CD71^{lo} cells from human bone marrow sorted as described in Fig. 1. The boxed area indicates the sort window that was used to recover CD34$^+$ CD71^{lo} cells from the cultures for subsequent cultures.

fluorescence of CD34$^+$CD71^{lo} cells is followed at various time intervals. The second approach involves sorting of single CD34$^+$CD45RA^{lo}CD71^{lo} cells and analysis of cell production in individual wells of culture plates. It is hoped that studies of cell populations combined with studies of individual cells will result in a better understanding of the mechanism by which primitive hematopoietic cells are maintained.

ACKNOWLEDGEMENTS

These studies were supported from grants from the NIH (AI29524) and the National Cancer Institute of Canada. Dr. M. Strong and colleagues from the North West Tissue Centre (Seattle, WA) are thanked for their help in making bone marrow cells from organ donors

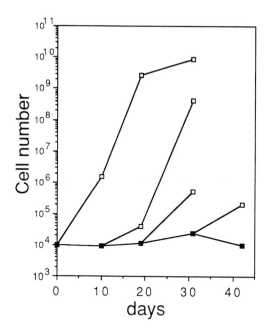

Figure 4. Maintenance of CD34$^+$CD71^{lo} (■-■) cells in rapidly proliferating serum-free suspension cultures containing IL-6, IL-3, MGF and Epo. Cultures were initiated with CD34$^+$CD45RA^{lo}CD71^{lo} cells from human bone marrow of organ donors (Fig. 1) and CD34$^+$CD71^{lo} cells were sorted from expanded cultures (Fig. 3) at the indicated time intervals. The total numbers of cells (□-□) present at each time point was calculated from cell counts of the (diluted) cultures. The total number of CD34$^+$CD71^{lo} cells in the cultures was calculated from the cell counts and the phenotypic analysis (Fig. 3). The actual number of CD34$^+$CD71^{lo} cells recovered after sorting was typically approximately 50% of this calculated number as a result of cell losses upon staining and sorting. Note the impressive proliferation of the cells (i.e. >10^6-fold increase in cell number on a per cell basis over a 30 day culture period) in these serum-free cultures.

available for these studies. Dr. D.E. Williams (Immunex, Seattle, WA) is thanked for gifts of recombinant growth factors. Colleagues from the Bone Marrow Transplantation Program of B.C. and the Terry Fox Laboratory are thanked for their help in making bone marrow samples, growth factors and tissue culture medium ingredients available for this study. The expert technical assistance of G. Thornbury, S. Abraham, C. McAloney and C. Smith is gratefully acknowledged.

REFERENCES

1. Lansdorp PM, Dragowska W: Long-term erythropoiesis from constant numbers of CD34$^+$ cells in serum-free cultures initiated with highly purified progenitor cells from human bone marrow. J. Exp. Med. 175: 1501-1509, 1992.
2. Lansdorp PM, Smith C, Safford M, et al: Single laser three color immunofluorescence staining procedures based on energy transfer between phycoery-thrin and cyanine 5. Cytometry 12:723-730, 1991.
3. Iscove NN, Guilbert LJ, Weyman C: Complete replacement of serum in primary cultures of erythro-poietin-dependent red cell precursors (CFU-E) by albumin, transferrin, iron, unsaturated fatty acid, lecithin and cholesterol. Exp. Cell Res. 126:121-126, 1980.
4. Sutherland HJ, Lansdorp PM, Henkelman DH, et al: Functional characterization of individual human hematopoietic stem cells cultured at limiting dilution on supportive marrow stromal layers. Proc. Natl. Acad. Sci. USA 87:3584-3588, 1990.
5. Horan PK, Slezak SE: Stable cell membrane labelling. Nature 340:167-168, 1989.

3

FLOW CYTOMETRIC PROLIFERATIVE FRACTION ANALYSIS IN SOLID
TUMORS

Daniel W. Visscher, Susan M. Wykes, and John D.
Crissman

INTRODUCTION AND BACKGROUND

Inappropriate growth regulation is a defining
feature of malignant neoplasia which, to a great extent,
becomes manifest as abnormal proliferation of neoplastic
populations. In recent years, it has become clear that
cell growth and proliferation are controlled by an
elaborate homeostatic mechanism which is balanced by
numerous intracellular and extracellular signals. It is
hardly surprising then that oncogenes and tumor suppressor genes which mediate this process have been found
to have great relevance in neoplastic transformation and
progression, reflecting clinical behavior. Recent
molecular genetic advances, however, reflect numerous
traditional observations which elegantly documented the
critical importance of abnormal growth regulation and
cell cycling (1). Prognosis in many clinical tumor
systems is strikingly correlated to arguably pedestrian
estimates of proliferation such as mitotic counts on
histologic tissue sections and empirical doubling times
extrapolated from serial radiographic measurements
(2,3).

The biological relevance of cell proliferation goes
well beyond mere determination of how rapidly a given
tumor will grow. There is considerable evidence to
suggest some chromosomal/genomic alterations result from
mitotic errors. In this context, the genetic instabil-

ity (or accumulation of genetic mutations) which characterizes malignant tumors may in part be driven by cell
cycling. Further, phenotypic properties requisite for
host invasion and/or metastasis, such as motility or
abnormal adhesion, may be coupled to cell cycling.
Finally, the effect(s) of various therapeutic interven-
tions is largely dependent on cell cycling characteristics of target populations. Clearly then, there is
considerable theoretical as well as practical utility to
studies of neoplastic cell proliferation.

The growth rate (or doubling time) of a neoplasm is
a complex function of multiple proliferation-related,
and unrelated, parameters - of which only a minority are
represented by any given measurement. It is important
to distinguish cell cycle distribution, or the relative
proportion of cells in G_0, G_1, S, G_2 and M, from cell
cycle kinetics, which is the length of time required for
a cell to complete the cell cycle. All static measurements, such as flow cytometric DNA analysis, assess the
former. The extent to which kinetics and cell cycle
distribution are related may vary between individual
tumors but both are determinants of growth rate. Moreover, the growth fraction of a population is determined
by the proportion of cells in G_1, S, G_2 and M. DNA
analysis cannot distinguish G_0 from G_1 events, and thus
proliferative activity assays based on DNA content
represent partial estimates of true growth fraction.

Before outlining the specific limitations and
artifacts of FCM proliferative fraction determinations,
it is perhaps worth mentioning that SPF is strongly
correlated with the other parameter routinely obtained
from DNA histograms - namely clonal DNA content, or
ploidy status (4,5). Although many clinical DNA analysis series omit SPF data, a growing number of retrospective studies employing multivariate statistical analysis
report that the prognostic significance of SPF outweighs

that of ploidy (5). These considerations imply DNA aneuploidy imparts aggressiveness largely, if not exclusively, by virtue of proliferation potential. It should also be mentioned that FCM provides rapid, quantitative and automated analysis of many thousands of cells. These features, as will be seen, offer significant theoretical and practical advantages over other technologies. Thus, the value of optimizing DNA cell cycle analysis for clinical and biologic studies of neoplasia is unquestionable.

TECHNICAL AND THEORETICAL LIMITATIONS

Any practitioner of FCM DNA analysis using clinical tumor specimens confronts a daunting list of obstacles to reliable SPF estimates (Table 1). Calculation of SPF, even from pristine unimodal cytophotometric histograms, presents a set of issues largely unresolved in the cytometry literature. DNA content represents a discrete function and, in theory, separation of G_0/G_1 from G_2/M and S cells is straightforward. All cytophotometric analyses however, represent DNA content as optical density or fluorescence intensity measurements. Perturbations caused by inconsistent dye binding, quenching and imperfect optics of detection systems (among other things) cause DNA histograms to consist of multiple, partially overlapped population distributions. In other words, signals derived from both G_0/G_1 and G_2/M cells overlap the S-phase region in DNA histograms. Similarly, S-phase event signals overlap G_0/G_1, and G_2/M regions. Thus, all FCM-derived cell-cycle determinations represent statistical, as well as biological, approximations of growth rate (see below).

SPF calculations, in essence, estimate the area under the S-region of the DNA histogram. Obviously, this requires knowing the channel numbers at which the S-region starts and ends, as well as how it is shaped.

Table 1

Limitations and Corresponding Approaches to
Optimizing FCM Cell Cycle Analysis

1.	Debris and aggregates	Mathematical subtraction algorithms Improved dissociation protocols
2.	Heterogeneity Small aneuploid population	Multisite sampling
3.	Population overlap - benign + neoplastic - multiple neoplastic stemlines	2-color analysis ?
4.	Interlaboratory variation - technique related - calculation related	Standardization and inter-laboratory norms

Validation of various mathematical SPF estimates by comparisons with another proliferative index (radiolabelled thymidine uptake) has been performed by Baisch (6) using cell lines (thereby avoiding tissue dissociation artifacts). At this point, it should be noted that _all_ measures of S-phase fraction are more or less subject to measurement artifacts and, therefore, no true SPF standard exists. Indices which directly assess incorporation of DNA metabolites, such as radiolabelled thymidine or bromodeoxyuridine, are analogous to SPF and represent a theoretical "gold standard" (7-9). Immunohistologic indices, such as ki-67 and PCNA staining, are less well correlated to SPF by virtue of expression in other cell cycle phases (10-13). Nevertheless, the studies of Baisch revealed that the "optimal" mathematical model depends on both the SPF level and the G_0/G_1 coefficient of variation (CV). For asynchronous populations having an SPF <30%, "true" proliferative fraction was best approximated by a rectangular model of SPF region shape.

Moreover, significant deviations of SPF became apparent as G_0/G_1 CV exceeded 5%.

An intuitive manual algorithm, which assumes the S-phase region is rectangular, extrapolates the S-phase fraction by sampling a contiguous representative segment of S-phase channels. In this model, the height of the S-phase rectangle consists of the average number of events in 10 selected channels clearly within the S-phase region. The base is the number of channels between the G_0/G_1, and G_2/M modes. Multiplying base and height thereby provides the S-phase integral. Once the S-phase events are calculated, G_0/G_1 and G_2/M are derived by correcting for _estimated_ overlap with the S-phase region. Clearly, the accuracy of SPF determination using this method is largely dependent on the G_0/G_1 CV, which determines the degree of overlap with the S-phase region, and the ability to define peak modes for G_0/G_1, and G_2/M. Although simplistic compared with other algorithms, we emphasize this method as it approximates SPF in biological systems well given low G_0/G_1 CV (<5%). This is a necessary requirement due to the need to identify the means and modes of G_0/G_1 and G_2/M populations. SPF estimates based on other assumptions about the shape of the S-phase region, including trapezoidal and polynomial distributions, have also been employed.

Observations confirming the validity of this approach have been made by authors who compared thymidine labelling index to FCM SPF of dissociated human tumors. Meyer and McDivitt reported FCM SPF correlated best with TLI measurements if the former was extrapolated from a contiguous 7-10 channel segment of the S-phase region containing the _fewest_ events (14). This strategy presumably eliminates debris (see below) as well as cells "arrested" in S-phase due to ischemia.

The concept of extrapolating SPF estimates from a

limited region of the histogram holds considerable utility in DNA aneuploid histograms, where the late S and G_2/M regions of diploid range populations overlap segments of the aneuploid S-phase region. In theory, cell cycle calculations in aneuploid DNA histograms should be possible using the previously described method unless the DNA index is peridiploid (i.e. 0.8-1.2), assuming "low" CV's. Although S-phase estimates derived from DNA aneuploid histograms with overlapping diploid range cell populations are a well known source of angst within the cytometry community, they have actually been shown to correlate better with TLI than than S-phase estimates from theoretically more straightforward diploid range histograms (7,15). This is because the DNA histogram of diploid range tumor cells is completely superimposed on histograms from "contaminating" stromal and inflammatory cells with a much lower SPF component. It should be recalled that, even in selected tissue sections of "pure neoplasm", approximately 40-60% of all cells are non-neoplastic. Events derived from these non-cycling cells disproportionately elevate G_0/G_1 peaks in diploid flow cytometric histograms, thus diluting the calculated S-phase. Despite such considerations, cell cycle data collected from diploid-range histograms are virtually always reported in clinical FCM studies. In contrast, SPF estimates in DNA aneuploid cases are not attempted in varying proportions of cases, depending on individual author (5) (see below).

SPF estimates in histograms with DNA aneuploid populations become especially problematic when the aneuploid clone is small (i.e. relative to the diploid range populations) or when multiple DNA aneuploid G_0/G_1 peaks are present. Although small DNA aneuploid stemlines may reflect a sampling artifact, the latter problem represents a technically insoluble issue arising in 8-12% of clinical solid tumor specimens. Most experts agree that SPF estimates are unreliable unless the aneuploid G_0/G_1 stemline constitutes at least 15% of histogram events (16-17). If more than one DNA stemline is present, manual approaches to cell cycle analysis are often not possible, since non-overlapped S-phase regions may not be present. It should also be noted that diploid range G_0/G_1 peaks are assumed to represent non-neoplastic populations by virtually all authors. However, there is accumulating evidence to indicate that cytometrically detectable diploid range clones are not infrequent in tumors with aneuploid stemlines (18). Further, there is no compelling reason to believe that they are biologically less relevant than co-existing aneuploid clones.

Clonal DNA content heterogeneity of neoplastic populations has implications more far reaching than mere DNA histogram complexity. Tumor heterogeneity is fundamental to the evolution of malignant neoplasms and is reflected in a wide spectrum of phenotypic traits, especially proliferation (19-20). Indeed, FCM studies employing multisite sampling protocols have documented impressive, if not disturbing, cell cycling heterogeneity (21). It may be argued that this simply documents technical inconsistencies of FCM analysis (22), however, these observations are confirmed in studies employing other cell proliferation assays (21). A critical issue, therefore, is whether any method reliably identifies the most aggressively growing subpopulation within a given neoplasm. We would parenthetically note that flow cytometry, by virtue of its abundant sampling capability, offers very real advantages over other technologies in confronting the issue of tumor heterogeneity.

Artifacts created by tissue dissociation represent another possibly more significant challenge to clinically meaningful cell cycle calculations. Nuclear debris is created by all tissue disaggregation methods, but

particularly by enucleation from paraffin-embedded
tissue blocks. Some nuclear fragmentation is accounted
for by nuclear slicing during section cutting, however
cytolysis due to autolysis, incomplete fixation and
enzyme digestion certainly represent additional causes.
In DNA histograms, debris is easily recognized as a
characteristic descending curve which partially overlaps
the diploid range G_0/G_1 population, and thereby widens
its CV. Much discussion in clinical cytometry literature is devoted to speculation about impaired resolution
of hyperdiploid peaks due to debris. Given the considerations outlined above, however, the effect(s) of
debris in cell cycle analysis are clearly profound.
Virtually all authors who address the issue of debris
subtraction, either technically or through empirical
mathematical models, report that debris accounts for up
to 50% of all "S-phase events" in uncorrected histograms
from paraffin-embedded specimens (7,23). This would not
constitute a major problem if the relative amount of
debris was consistent from case to case. As all users
of flow cytometric DNA analysis have discovered though,
intertumoral variability in histogram debris may be
profound.

Another dissociation-related parameter is the
presence of nuclear aggregates (i.e. doublets and triplets and fragments of "sliced" nuclei) which form in
nuclear suspensions due to the inherent "stickiness" of
DNA. Aggregates of intact nuclei are easily recognized
as discrete populations in 2-parameter dot plots (size
vs DNA content), which interfere with cell cycle calculation by virtue of overlapping the diploid range G_2/M
peak (doublets) or aneuploid S-phase region (triplets).
Aggregates of intact and partial nuclei will distribute
throughout the histogram and essentially appear to
represent debris.

The final dissociation-related parameter is repre-

sentative of neoplastic populations, even in carefully
sampled tissue slices. A growing body of evidence
suggests highly aneuploid and/or rapidly cycling cells
are disproportionately susceptible to cytolysis during
disaggregation (24). This may account for the relative
paucity of hypertetraploid events/populations in FCM DNA
analysis series compared to some image analysis studies.

Interestingly, the most significant limitation to
the utility of FCM DNA analysis may be philosophical
rather than technical or biological. Both disaggregation procedures and histogram analysis algorithms vary
widely in the published cytometry literature. This is
reflected in the striking variability between mean SPF
values in representative series from a given tumor
system (5). More telling perhaps, are disconcerting
interseries differences in the percent of cases for
which cell cycle data are reported. Thus, even among
the "experts" there are major disagreements about when
SPF estimates are appropriate.

Solutions to the Quandary

An ideal FCM cell cycle analysis method would meet
several important criteria. Most importantly, it would
have to produce an accurate reflection of biologic
reality, or "true" neoplastic proliferative fraction.
Validation studies through comparisons to independent,
but analogous, cell proliferation assays or clinical
parameters would thus be necessary. Second, it would
need to be applicable, with similar accuracy, to the
vast majority of DNA histogram types. Third, it would
have a high level of intra-user and inter-user reproducibility. In other words, it could be standardized.
Finally, it would come with unequivocal criteria for
histograms which are unacceptable, or too complex, for
reliable analysis. Given the seriousness of the technical, biological and philosophical issues already dis-

cussed, there is no obvious solution to this problem. The problem, moreover, involves the whole of FCM DNA analysis methodology and interpretation and not just cell cycle calculations per se. Solutions which have been proposed in the published literature, in general, are partial. That is, they attack specific limitations individually, rather than attempting a comprehensive restructuring of cell cycle analysis. These partial solutions are listed in Table 1.

Mathematical Modeling

A number of commercially-available software programs perform automated DNA cell cycle analysis using a technique called mathematical modeling. In essence, these "curve-fitting" or "deconvolution" programs employ non-linear least squares analysis to "fit" collected data with predicted shapes of cell cycle histograms - namely a series of Gaussian (normal) distributions (17). Despite the statistical elegance and visually pleasing data reductions offered by these programs, it must be emphasized that they cannot perform accurate cell cycle analysis of complex histograms unless some portions of S and G_2/M are non-overlapping. Although largely dependent on DNA index and CV, this situation pertains primarily in cases with near-diploid DNA content (i.e. DNA index 0.8-1.2). It will be recalled that this is precisely the situation in which manual cell cycle calculation was impossible. Therefore, mathematical modeling algorithms do not necessarily extract more relevant clinical or biological data from DNA histograms than manual techniques. Moreover, their efficacy is limited by histogram quality (i.e. CV). A principle advantage of modeling software, it would seem, resides in the consistency and objectivity with which a computer program analyzes data, providing a theoretical basis for standardization (at least in histogram analysis). The

relevance of cell proliferation data collected by automated mathematical modeling algorithms has been demonstrated in retrospective clinical studies (25). We are
unaware of studies which directly compare the clinical
predictive value of automated and manual SPF calculation
techniques.

Debris and Aggregate Subtraction

One area in which histogram data manipulation has
proven benefit in optimizing the clinical relevance of
FCM is mathematical or signal-based subtraction of
debris and aggregates. In a sense, nuclear debris and
aggregation represent opposite sides of the same coin -
the former representing fragmentation caused by slicing
or lysis and the latter representing re-association of
once disaggregated intact, or partial, nuclei. The
extent to which these phenomena occur is highly dependent on specimen type (i.e. fresh/frozen vs paraffin-
embedded), fixation, and dispersion technique (mechanical vs enzymatic). However, even with similar protocols, the relative amount of debris may vary dramatically between individual cases. Both are especially
prominent following enucleation of paraffin-embedded
samples and, as previously noted, result in visually
obvious histogram artifacts. Before proceeding to a
description of debris or aggregate compensation, it is
worth mentioning that the true distribution of either
contaminant within a given DNA histogram is not known.
All debris or aggregate subtraction algorithms thereby
combine actual histogram data with assumptions about the
expected distribution to extrapolate the amount of
either component.

Early debris-subtraction methods simply "fit" the
observed debris slope on the left of G_0/G_1 to an exponential function which was then extrapolated into the S-
phase region of the histogram. This formula may over-

compensate for debris in the S-phase region, resulting in zero or negative synthesis phase fraction values in some cases. The weakness of exponential curve fitting was that it didn't account for a very reasonable assumption concerning debris origin and distribution - namely that debris signals are proportional to the number of nuclei in a given population. Since the G_0/G_1 population contains the largest number of cells, it would be expected to produce the most debris and thereby have the "highest" debris slope. Similarly, the smaller number of S and G_2/M cells would be expected to produce a proportionately "lower" debris curve. More recent debris subtraction algorithms, which adjust debris subtraction for magnitude and positioning of histogram populations, produce debris slopes characterized by a sharp step-off when the curve approaches the middle of the G_0/G_1 peak (i.e. as the debris from this population is "left behind") (16,17). This approach to debris subtraction assumes that there are no partial nuclear aggregates.

The clinical utility of histogram debris subtraction, at least for studies employing paraffin enucleation, is now well established. Various authors have reported improved correlations of debris subtracted SPF to other methods of proliferative fraction assessment (such as TLI) (24,26), clinicopathologic parameters (27), and even disease outcome (25). Widely utilized commercially-available cell-cycle analysis software now routinely employ debris subtraction.

The other side of the coin, aggregates, interfere with cell cycle analyses primarily by contaminating the diploid range G_2/M population (i.e. doublets) or the aneuploid S-phase region (triplets). Theoretically, aggregates may be formed by any combination of elements in suspension, including various combinations of diploid and aneuploid nuclei. Debris or partial nuclei may also form aggregates with one another or with intact nuclei. Mathematical aggregate compensation theory, therefore, is potentially quite complex (16). To a great extent, aggregate contamination may be controlled by use of careful dissociation technique.

Two approaches have been applied to aggregate subtraction from DNA histograms. The first employs electronic gating based on detection of pulse-shape differences between singlet and doublet nuclei. Basically, doublets may produce a bimodal pulse vs time signal as they transverse the aperture due to their dumbbell-like shape. In contrast, singlets produce a unimodal signal. It is now possible to purchase software (Becton-Dickinson) which excludes aggregates based on altered pulse shape. Unfortunately, pulse shape is also dependent on nuclear shape as well as the alignment of aggregates as they pass the laser. Moreover, triplets are difficult to detect with this method due to less distortion of pulse shape. Accordingly, some studies report pulse height discrimination results in partial aggregate compensation (28).

The other approach involves mathematical modeling algorithms which formulate the expected distribution of aggregate peaks based on statistical estimates of interaction likelihood between various species entered into the model. These software programs establish the degree of aggregate formation within a given specimen by manually assessing the presumed triplet population. The content of doublets "hidden" under G_2/M is then calculated by a mathematical model based on the expected statistical relationship to triplets. Mathematical aggregate compensation produces significant modifications of S-phase estimates which have been shown to correlate with microscopic assessments of aggregates (16). Utility in clinical follow-up series, however, has yet to be fully established.

Multiparametric Analysis

Optimization of cell cycle calculations discussed until this point have been largely cosmetic. They have not addressed the real issue - presence of overlapping cell cycle distributions from different, biologically distinct, populations. It will be recalled that, although S-phase estimates in diploid range histograms are mathematically more straightforward than in DNA aneuploid histograms, they are biologically less representative since histograms of neoplastic and benign populations overlap completely. In order to achieve our first criterion of "ideal" cell cycle analysis, then, we must find a way to make DNA histograms more representative of neoplastic elements. One such approach, which has been studied and advocated by our group, is 2-color, multiparametric analysis. In this technique, intact cells are labelled with fluorescent-conjugated antibodies to cytoplasmic constituents. Using computer gating, these populations may than be isolated for cell cycle analysis.

We have employed 2-color analysis with antibodies to cell lineage specific antigens, such as cytokeratin or leukocyte common antigen (26). There is no theoretical reason, however, why these studies could not be extended to tumor-specific antigens, such as oncogene products. Figure 1 outlines the DNA histograms produced by "gating" on cytoplasmic constituents in a dissociated breast carcinoma. As can be seen, the problem of overlapped cell cycle distribution in this peridiploid (DI=1.2) tumor is eliminated by "gating-out" non-neoplastic inflammatory and stromal cells. SPF calculations using this method are precluded only in tumors which contain multiple DNA stemlines.

Whole cell FCM analysis has the additional advantage of allowing less rigorous mechanical tissue disaggregation, resulting in minimal debris. In addition,

Figure 1.   Two-color multiparametric DNA analysis of a breast carcinoma.

Top left: Ungated DNA histogram, showing two G_0/G_1 peaks, the first at channel 50 and a second at channel 100. Note the G_2M (channel 200) which corresponds to the second population.

Top right: Histogram of green-fluorescence negative control tube, consisting of suspension incubated with FITC-conjugated non-immune antibody. The cursor (channel 85) denotes "gate" used to exclude fluorescence caused by non-specific binding/auto-fluorescence.

Middle left: Green-fluorescence histogram of suspension "labelled" with FITC-conjugated anti-leukocyte common antigen. Note position of "gate" (channel 85) and population corresponding to tumor-infiltrating inflammatory cells.

Middle right: DNA histogram of LCA-positive cells. This establishes channel 50 as the fluorescence intensity of diploid range events. Note the low synthesis phase fraction of this population.

Bottom left: Green-fluorescence histogram of suspension "labelled" with FITC-conjugated anti-cytokeratin. As expected, there are considerably more epithelial than inflammatory cells in this dissociated carcinoma.

Bottom right: DNA histogram of cytokeratin-positive cells, showing a diploid-range population (channel 50) and a tetraploid population (channel 100) which has been enriched by gating on epithelial cells (compare with top left). The diploid range events may represent a neoplastic population, however "contamination" from residual benign epithelium is not excluded since cytokeratin is cell-lineage, but not tumor cell, specific.

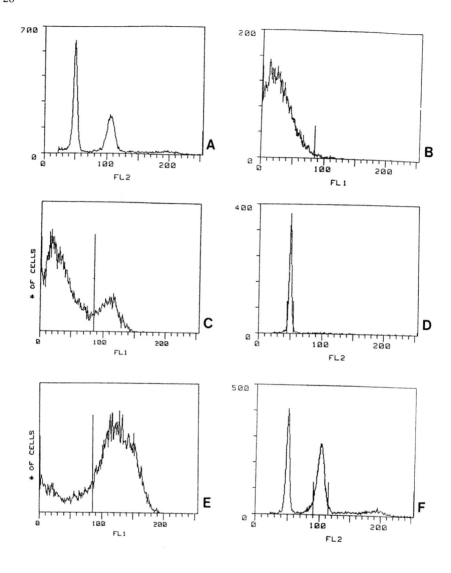

aggregates are less likely to form with intact cells. The only significant disadvantage to this approach is the absolute requirement for fresh, unfixed tissue which mandates prospective studies. Obviously, not all flow cytometry facilities will have a source of fresh tissue aliquots. For these reasons, few clinical studies employing 2-color analysis have been published.

REPORTING

One of the most troublesome issues in clinical DNA analysis is how cell cycle data should be reported. Obviously, most clinicians lack appreciation of what constitutes "low" vs "high" SPF. Further, given the already noted interlaboratory variabilities, there is little point in adopting the cut-off values generated by other facilities. We recommend that each lab establish internal median SPF values for specific tumor types and report individual cases in reference to those figures. Prognostic studies in the literature have employed various methods of comparing SPF with outcome. Most authors analyze SPF as above vs below median, however some use quartiles or even examine SPF as a continuous variable. The optimal, or most useful, method of analyzing SPF (or any proliferation-related index) is not yet established.

EPILOGUE

The accuracy and reproducibility of flow cytometric cell cycle analysis, therefore, is limited by technical, biological and philosophical problems. For the everyday user of FCM DNA analysis, these issues translate into predictable histogram artifacts, including, impaired peak resolution (high CV), incomplete specificity (tumor cell histograms superimposed on normal cell, debris or aggregate events), inadequate representation and heterogeneity (small or multiple aneuploid populations) as well as confusion about histogram analysis. We have shown that, as the causes of these problems are multifactorial, so are the presently-available solutions. Varied approaches to these issues, in large part, accounts for the conflicting, inconsistent cell cycle data published in the cytometry literature.

As a developing technology, it is hardly surprising that FCM DNA analysis has been subject to methodologic and philosophical inconsistencies. Much discussion has centered on interlaboratory comparison studies and standardization consensus meetings, ostensibly in order to further the role of DNA analysis in clinical patient management. Despite the uncontested prognostic relevance of growth rate assessment, however, it is not entirely clear whether even the most accurate FCM ploidy and S-phase determinations will gain widespread acceptance as clinically useful parameters. The issue of optimizing FCM cell cycle analysis, in our opinion, also reaches well beyond "fine-tuning" of current methodologies for clinical specimens, into development of creative new approaches to cytometric study of neoplastic cells. Clearly, this involves concurrent advances in related fields, such as proliferation markers and interphase cytogenetics, as well as badly needed re-assessment of matters such as fixation and dissociation. Given the near universal availability of flow cytometers, we anticipate continued evolution and more accurate assessment of cell cycle analysis to continue.

REFERENCES

1. Malaise EP, Chavaudra N, Charbit A, Tubiana M: Relationship between the growth rate of human metastases, survival and pathologic type. Eur. J. Cancer 10:305-312, 1974.
2. Pearlman AW: Breast Cancer. Influence of growth rate on prognosis and treatment evaluation. A study on mastectomy scar recurrences. Cancer 38: 1826-1833, 1976.
3. Tubiana M, Koscielny S: Cell kinetics, growth rate and the natural history of breast cancer. Eur. J. Cancer Clin. Oncol. 24:9-14, 1988.
4. McDivitt RW, Stone KR, Craig B, et al: A proposed classification of breast cancer based on kinetic information. Derived from a comparison of risk factors in 168 primary operable breast cancers. Cancer 57:269-276, 1986.
5. Visscher DW, Sarkar FH, Crissman JD: Clinical significance of pathologic, cytometric and molecular parameters in carcinoma of the breast. In: Advances in Pathology and Laboratory Medicine, RS

Weinstein (ed), Mosby, Vol. 5, 1992.

6. Baisch H, Beck HP, Christensen IJ, et al: A comparison of mathematical methods for the analysis of DNA histograms obtained by flow cytometry. Cell Tissue Kinet. 15:235-249, 1982.

7. Meyer JS, Coplin MD: Thymidine labeling index, flow cytometric S-phase measurement, and DNA index in human tumors. Am. J. Clin. Pathol. 89:589-595, 1988.

8. Waldman FM, Chew K, Ljung BM, et al: A comparison between Bromodeoxyuridine and ^{3}H thymidine labeling in human breast tumors. Modern Pathology 4:718-722, 1991.

9. McDivitt RW, Stone KR, Meyer JS: A method of dissociation of viable human breast cancer cells that produces flow cytometric kinetic information similar to that obtained by thymidine labeling. Cancer Res. 44:2628-2633, 1984.

10. Garcia RL, Coltrera MD, Gown AM: Analysis of proliferative grade using anti-PCNA/cyclin monoclonal antibodies in fixed, embedded tissues. Am. J. Pathol. 134:733-739, 1989.

11. Sahin AA, Ro JY, El-Naggar AK, et al: Tumor proliferative fraction in solid malignant neoplasms. A comparative study of Ki-67 immunostaining and flow cytometric determinations. Am. J. Clin. Pathol. 96:512-519, 1991.

12. Dawson AE, Norton JA, Weinberg DS: Comparative assessment of proliferation and DNA content in breast carcinoma by image analysis and flow cytometry. Am. J. Pathol. 136:1115-1124, 1990.

13. Vielh P, Chevillard S, Mosseri V, et al: Ki67 index and S-phase fraction in human breast carcinomas. Am. J. Clin. Pathol. 94:681-686, 1990.

14. McDivitt RW, Stone KR, Craig RB, Meyer JS: A comparison of human breast cancer cell kinetics measured by flow cytometry and thymidine labeling. Lab. Invest. 52:287-291, 1985.

15. Isola JJ, Helin HJ, Helle MJ, Kallioniemi OP: Evaluation of cell proliferation in breast carcinoma. Comparison of Ki-67 immunohistochemical study, DNA flow cytometric analysis, and mitotic count. Cancer 65:1180-1184, 1990.

16. Rabinovitch PS: Practical considerations for DNA content and cell cycle analysis. In: Clinical Flow Cytometry, KD Bauer, RE Duque, TV Sharkey (eds.), Williams and Wilkins, Baltimore, pp. 117-142, 1990.

17. Bagwell CB: Theoretical aspects of flow cytometry data analysis. In: Clinical Flow Cytometry, KD Bauer, RE Duque, TV Sharkey (eds.), Williams and Wilkins, Baltimore, pp. 42-62, 1990.

18. Beerman H, Smit VT, Kluin PM, et al: Flow cytometric analysis of DNA stemline heterogeneity in primary and metastatic breast cancer. Cytometry 12:

147-154, 1991.

19. Olszewski W, Darzynkiewicz Z, Rosen PP, et al:
Flow cytometry of breast carcinoma III. Possible
altered kinetics in axillary lymph node metastases.
Analy. Ouant. Cytol. 4:275-278, 1982.

20. Nicholson GL: Tumor cell instability, diversifica-
tion, and progression to metastatic phenotype: From
oncogene to oncofetal expression. Cancer Res.
47:1473-1487, 1987.

21. Meyer JS, Wittliff JL: Regional heterogeneity in
breast carcinoma: Thymidine labelling index, steroid hormone receptors, DNA ploidy. Intl. J. Cancer 47:213-220, 1991.

22. Kute TE, Gregory B, Galleshaw J, et al: How reproducible are flow cytometry data from paraffin-embedded blocks? Cytometry 9:494-498, 1988.

23. Feichter G, Czech W, Haag D, et al: Comparison of
S-phase fractions measured by flow cytometry and
autoradiography in human transplant tumors. Cytometry 9:605-611, 1988.

24. Weaver DL, Bagwell CB, Hitchcox SA, et al: Improved flow cytometric determination of proliferative activity (S-phase fraction) from paraffin-
embedded tissue. Am. J. Clin. Pathol. 94:576-584,
1990.

25. Rabinovitch PS, Barlow W, Visakorpi T, et al:
Improving the prognostic strength of S and G2 estimates by optimizing DNA histogram analysis in
breast cancer. Cytometry (suppl.) 6:82, 1993.

26. Haag D, Feichter G, Goerttler K, Kaufmann M: Influence of systematic errors on the evaluation of
the S-phase portions from DNA distributions of
solid tumors as shown for 328 breast carcinomas.
Cytometry 8:377-385, 1987.

27. Visscher DW, Zarbo RJ, Jacobsen G, et al: Multiparametric deoxyribonucleic acid and cell cycle
analysis of breast carcinomas by flow cytometry.
Clinicopathologic correlations. Lab. Invest. 62:
370-378, 1990.

28. Seamer LC, Bunt JC, Dressler LG, Revels A: A comparison of pulse gating and mathematical modeling
methods of reducing the contribution of aggregates
to flow cytometric DNA single parameter histograms.
Cytometry (suppl.) 6:44, 1993.

4

MITOGENIC AND NON-MITOGENIC INDUCTION OF LYMPHOCYTIC
INVASION: DUAL PARAMETER FLOW CYTOMETRIC ANALYSIS

S. Ratner and D. Lichlyter

INTRODUCTION

It has been well established that mitogenic activation produces profound alterations in the migratory
behavior of lymphocytes. Activated, blastic T and B
cells down-regulate homing receptors which mediate
adhesion to the high endothelium of peripheral lymph
nodes (1-3). Concurrently, they up-regulate adhesion
receptors for extracellular matrix components and for
ligands typical of inflammatory endothelium (4-6). The
result is a change in traffic patterns, from extravasation into lymph nodes to extravasation into nonlymphoid
tissue, especially inflamed tissue (7,8). After the
immune response terminates, memory cells, the long-lived
residual progeny of the blasts, remain. Memory helper T
cells, even when not proliferating, retain the non-
homing, inflammation-seeking behavior of the parental
blasts (9-11). There are indications that other memory
lymphocyte subsets follow similar trends (12,13).

We and others have found that mitogen-induced
changes in traffic pattern are mirrored in the motile
response of lymphocytes to extracellular matrix (ECM) in
vitro. Mitogenic stimuli induce in murine and human B,
T, and NK cells the capacity to respond chemotactically
to soluble fibronectin and laminin (14); and, to adhere
to and migrate on coatings of fibronectin, Type I collagen, and laminin (15-17). Also induced is the ability

to invade fibrous gel lattices of Type I collagen, collagen-fibronectin alloy, and Matrigel basement membrane matrix (18-21). Recently, we have found that the invasiveness of murine T cells does not develop until proliferation is well under way, 48-72 h after the start of exposure to mitogens such as high-titer IL-2, anti-CD3 and PHA (21,22). Once induced, invasiveness is persistent. When cultures are maintained without further stimulation for 3-4 weeks, the surviving cells, though nonproliferative, migrate as avidly as blasts into ECM gels (22). These findings suggest that long-lasting reprogramming for migratory response to ECM is a component of the memory-cell phenotype.

Mitogen-induced activation cannot, however, be the only means of inducing lymphocyte motility. Naive, unstimulated lymphocytes, in the course of normal traffic, repeatedly leave the circulation, extravasating into lymphoid tissue and, at lower frequency, into nonlymphoid tissue. They then re-enter the circulation (1,7). This demands rapid and reversible transitions of behavior towards ECM, from nonadherence to adherence and migration.

Thus, there may be two routes through which lymphocytes gain the capacity for motility in ECM. Mitogenic induction, which stimulates delayed but long-lasting invasiveness, is easily produced _in vitro_, and its mechanisms can therefore be readily studied. Nonmitogenic induction, causing rapid and transient invasiveness in resting lymphocytes, is more difficult to model _in vitro_. The physiological triggers are still unknown. Likely candidates include contact with endothelial ligands and tissue-derived chemotactic factors. We have thus far been unable to induce prompt invasiveness in lymphocytes _in vitro_ through exposure to cultured endothelium and a panel of soluble chemotaxins. As the search for invasion-inducing receptors continues, we have been investigating the biochemical cascades which might transduce signals from such receptors. Activation of C-type protein kinases (PKC) is known to induce motility in many cell types (23-26), including lympho-cytes (27-29). We have recently determined that a brief pulse of the phorbol ester PMA induced lymphocyte inva-siveness into ECM in a rapid and reversible manner. PMA-induced invasion was nonmitogenic, being accompanied by negligible thymidine incorporation (22). Induction, then must occur prior to S phase.

This still leaves uncertainty as to whether PMA-stimulated lymphocytes must transit through G_1, before invasiveness is expressed. Resolution of this matter is an important initial step toward analysis of motility regulation at the molecular level. A widely used form of cell-cycle analysis, single-parameter propidium iodide fluorescence, is inadequate for the purpose, for it provides no resolution of G_0 and G_1 phases. In the present study, we employed acridine orange metachromatic fluorescence to analyze simultaneously the DNA and RNA content of lymphocytes. This was supplemented by immu-nofluorescence analysis of cell-surface molecules asso-ciated with G_1. We now report that PMA-induced inva-siveness appears at the onset of G_1.

MATERIALS AND METHODS
Lymphocyte Culture

Cells harvested from the peripheral lymph nodes of BALB/c mice were enriched for T cells (>90% CD3[+]) by passage through nylon-wool columns or, in some cases, were used unfractionated (approx. 70% CD3[+], 25% SIg[+]). The cells were washed, resuspended in enriched Dulbe-cco's Modified Eagle's Medium (DME) containing 5% calf serum and 5% fetal calf serum (FCS) at 6×10^6/ml. PMA dissolved in DMSO was added to a final concentration of 10 ng/ml. Control cultures received only DMSO. It was determined that DMSO at the concentration used, 0.001%, had no effect on motility. Highly proliferative lymphocyte populations were required as positive controls and compensation standards for 2-color analysis. For this purpose, lymph node lymphocytes were exposed to solid-state hamster anti-mouse CD3-epsilon (clone 145-2C11) for 2 days, then cultured in the presence of IL-2 (10-20 U/ml) for 1-3 additional days.

Lymphocyte Invasion System

The system has been described previously (21). Briefly, ice-cold liquid ECM components were poured into culture wells 13 mm in diameter and allowed to polymerize into a gel 4 mm thick. In the experiments reported here, the ECM was Type I collagen extracted from rat tail tendons and reconstituted with concentrated RPMI 1640 medium to physiological osmolarity and 1.2 mg collagen per ml gel. The gels contained either 10% FCS or 0.05% bovine serum albumin. The same proteins were present in the DME used to pre-equilibrate the gels prior to experiments. Lymphocytes suspended in DME were allowed to settle onto gel surfaces at 2.5 or 3.0×10^6 per well, or approximately 2×10^6 per cm^2 gel surface. Gels were incubated at $37^0 C$, usually for 6 h, then fixed in 1% paraformaldehyde for microscopic measurement of distance travelled by leading cell fronts. For separation of lymphocytes exhibiting invasion during the incubation period, the system was scaled up to employ gels 35 mm in diameter. Incubation lasted from hours 2 through 6 after the PMA or vehicle pulse, and the gels were not fixed. Nonadherent, surface-adherent, and motile (invasive) lymphocytes were fractionated by a standardized sequence of washes and collagenase treatments (21). In some wells, the three fractions were recombined to reconstitute the total population. Total populations and motile fractions were immediately fixed for cell cycle analysis.

Cell Cycle Analysis

A modification of the technique of Darzynkiewicz
(30) was employed. Cells were washed twice with serum-
free PBS and aliquoted into ice-cold PBS at $10^6/0.15$ ml.
Ice-cold absolute ethanol was added to a final concentration of 70%. After 20 min fixation with frequent
agitation, 4.5 ml PBS were added and the cells were
pelleted and resuspended in 0.2 ml PBS. For aliquots to
be used as red-channel compensation controls, the PBS
was supplemented with RNAse (Sigma, 65 Kunitz U/mg, 0.1
mg/ml) and incubated 20 min at 37°. Aliquots were again
chilled and each received 0.4 ml ice-cold Solution A
(0.15 N NaCl containing 0.08 N HCl and 0.1% Triton X-
100) and, 15 sec later, 1.2 ml ice-cold Solution B
(phosphate-citric acid buffer, pH 6.0 containing 0.15 N
NaCl, 1 mM EDTA, and 20 μM acridine orange [Harleco,
Philadelphia, PA]). Fluorescence was analyzed 3-15 min
later with a Becton-Dickinson FACStar flow cytometer
using 488 nm excitation. Live gating was used to exclude nonviable and nonlymphoid cells. Linear green
fluorescence (DNA) and red fluorescence (RNA) were
collected, respectively, with band-pass filters 530/30
nm and 660/24 nm. Crossover of green emission into the
red channel was compensated using RNAse-treated cells,
which emitted only green fluorescence. We were unable
to produce red-only controls by extensive treatment of
lymphocytes with DNAse; significant green fluorescence
always remained. An alternate method was therefore used
to compensate for red crossover into the green channel.
After red-channel compensation, anti-CD3-stimulated
lymphocytes were stained with acridine orange and analyzed. Compensation was adjusted so that the G_1 population, known to contain uniform DNA content, became
parallel to the green axis (Fig. 1B). A technical prob-

lem associated with acridine orange is its absorption by sample lines and leaching into subsequent samples. The problem was solved by the use of a separate sample line for acridine orange experiments, coupled with flushing of all lines with 10% household bleach.

Immunofluorescence Staining

Cells were stained with either rat-anti-mouse IL-2R (alpha-chain-specific, clone 7D4, FITC conjugate, Pharmingen, San Diego, CA) or rat-anti-mouse LECAM-1 (clone MEL-14) with secondary labelling by FITC-coupled anti-rat Ig $F(ab')_2$ fragment, Fc-specific, Jackson Labs, West Grove, PA). Flow cytometric analysis was as described above, except for collection of emission on a logarithmic scale.

RESULTS

Development of PMA-Stimulated Motility

A two-hour exposure to 10 nm PMA induced significant lymphocyte motility in collagen gels over the ensuing 6 h, both in terms of percentage of cells exhibiting motility and the distance travelled by the leading cell front (Table 1). T cells of the CD4+ and CD8+ subsets, as well as B cells, all showed comparable motility (not shown). Motility was significantly elevated over that of vehicle-treated controls for 6-16 h after the end of the PMA pulse; after this point it declined to control levels (Table 1). Analysis of the cell-cycle profiles of total populations and motile fractions was generally performed 4-6 h after the PMA pulse.

Cell-Cycle Analysis

Fresh, unstimulated lymph node lymphocytes from untreated animals can be assumed to be almost exclusively in G_0. They were therefore used as a reference popu-

Table 1

Development and Decline of PMA-Induced Motility in
Murine T Lymphocytes. Results of a Typical Experiment

Start of 6-h Assay (Hours after stimulation)	Motility			
	Leading-Front Distance[1]		Percent Motile[2]	
	vehicle	PMA	vehicle	PMA
0	22 (4)	144 (13)	1.3	8.2
4	22 (5)	152 (9)	ND[3]	ND
10	51 (3)	116 (2)	ND	ND
20	58 (8)	80 (7)	2.6	2.4

[1]Mean (μm $\pm$ S.D.) from four 100X fields in each of 3 replicate gels.
[2]From counts of motile fractions and total populations obtained as in Methods.
[3]Not determined.

lation to define basal levels of DNA and RNA (Fig. 1A).
For comparison, the cell cycle profile of a typical
activated T cell population is shown in Fig. 1B. This
population was stimulated for two days with solid-state
anti-CD3 and cultured in the presence of IL-2 (20 U/ml)
for an additional day. This is the point at which the
population first achieved maximal mitogen-induced motility (22). Only 15% of the population remained within
the G_0 boundary; 45% entered G_1, a phase defined by
increased RNA content without increased DNA. Cells in S
phase, detected by their increased RNA and DNA content
between N and 2N, accounted for 23% of the population.
The remaining 17% were in G_2M, defined by increased RNA
and approximately 2N DNA.

The cell cycle profile of the motile (invasive)
fraction of a typical PMA-stimulated population is
displayed in Fig. 1C. This fraction consisted of cells
which were liberated from beneath the surface of collagen gel after a migration period that began immediately

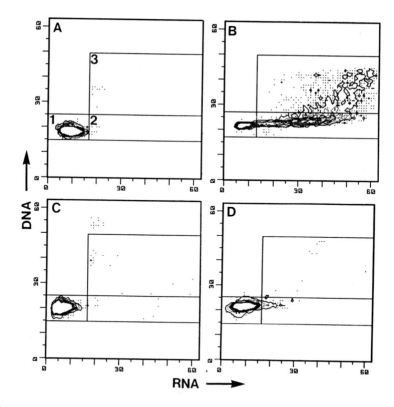

Figure 1. Dual parameter cell-cycle analysis by acridine orange metachromatic fluorescence. A. Fresh, unstimulated murine T cells; B. Proliferating T cell population, 3 days after start of stimulation with anti-CD3; C. Motile (invasive) fraction of PMA-stimulated population, collected after 4 h migration period in Type I collagen gel; D. PMA-stimulated T cell population, 24 h after PMA pulse. Analysis windows defining cell cycle phases are denoted in A as follows: $1=G_0$; $2=G_1$; $3=S$ and G_2+M.

after a PMA pulse and ended 4 h later. The cell cycle profile is indistinguishable from that of a fresh, resting population (Fig. 1A).

The acridine orange technique has been reported to resolve G_1 into subphases, including T, A, and B (30). We, however, were unable to resolve them, making it difficult to distinguish cells in G_0/G_{1Q} (noncycling) from those in G_{1T} (transition to G_1,) and G_{1A} (early G_1). Our fluorometric analysis therefore left open the possibility that motility induction was associated with early G_1. To test this idea, we assayed cell-surface changes known to be associated with G_{1A}: Increased expression of IL-2 receptors (31) and loss of the homing receptor for peripheral lymph node endothelium, gp90[MEL-14]. Immunofluorescence analysis demonstrated that these changes had indeed occurred in nearly 100% of the cells by the time peak motility had developed (8 h post-pulse, Figs. 2A and 2B). It can therefore be concluded that the PMA-stimulated acquisition of locomotory capacity is associated with entry into G_1, but so early in G_1 that RNA content had not increased to fluorimetrically detectable levels.

After PMA-induced motility declined, little further progression through cell cycle was detected by acridine orange fluorescence analysis. At 24 h post-pulse, less than 5% of the cells had entered G_1, and none had progressed to S (not shown). Expression of homing receptor gp90[MEL-14] returned to its original level (not shown), but IL-2R remained elevated, suggesting that the cells remained in an early phase of G_1.

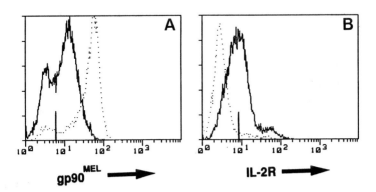

Figure 2. Immunofluorescence analysis of PMA-pulsed T cell population during the period of maximum invasiveness, 8 h after pulse (solid line); and of vehicle-treated control (dotted line). A. Expression of peripheral lymph node homing receptor. B. Expression of alpha chain of IL-2 receptor. Vertical marks denote maximal fluorescence of normal Ig-stained controls.

DISCUSSION

The induction of lymphocyte locomotory capacity by antigens or other mitogens is part of the process which mobilizes lymphocytes for immune response. Some signs of induction appear within 24 h, e.g. polarization and migration in response to soluable chemotactic and chemo-kinetic factors (18). Wilkinson and coworkers have obtained evidence that these changes are associated with passage through G_1. In mitogenically-stimulated human T and B cell populations, for example, the most polarized cells were those increasing in size and incorporating uridine but not thymidine (32). Ability to polarize failed to develop when mitogen treatment was performed in the presence of cyclosporin or FK506, inhibitors of G_1-related gene expression (33). These findings all suggest that mitogenically-stimulated lymphocytes develop polarization capacity and motility on artificial substrates as they progress through G_1.

Other manifestations of mitogen-induced motility take longer to develop. This includes the tendency to penetrate gels of reconstituted ECM, which may be an in vitro indicator of ability to invade tissue (34, and in preparation). We have found, for example, that 48-72 h elapsed between the start of mitogenic stimulation and the commencement of invasion into Type I collagen or Matrigel. This was the case for unseparated peripheral lymphocyte populations stimulated with high-titer IL-2 or the lectin PHA, and for T cell populations stimulated with anti-CD3 (21,22). Invasion into collagen was increased by the presence of fibronectin, but this response, too, took 48-72 h to develop. By this time, DNA synthesis and proliferation were well under way, with many cells completing their first passage through cell cycle (see Fig. 1B). We have also found that the motility-inducing effects of mitogenesis lasted well beyond the end of proliferation, and may in fact represent a permanent change in the behavioral program of the long-lived cells (22).

The present report supports the concept that lymphocytic invasion into ECM may also be induced rapidly, transiently, and without mitogensis. In this case, invasiveness was induced by brief stimulation with phorbol ester. This strategy will make it possible to study lymphocytic invasion mechanisms against a background uncomplicated by the expression of genes related to proliferation and mitogen-induced differentiation. It should also lead to insights into the mechanisms which allow naive lymphocytes to extravasate during the course of their normal traffic.

The invasion-inducing effects of PMA are attributable to its PKC-activating properties. Induction was abolished by PKC inhibitors such as H-7 and staurosporine (22, and in preparation). There was no evidence for the participation of tyrosine kinases, pertussis-toxin-sensitive G proteins, or calcium flux.

A crucial event in the induction of motility by PMA must be the up-regulation of expression or affinity of a Type I collagen receptor. The nature of this receptor is under investigation. We have so far determined that it is magnesium dependent but does not bind via RGD-containing sequences. This behavior is characteristic of a known human Type I collagen receptor, VLA-2 (35).

PMA-induced motility is not identical in all respects to that induced by mitogens. Mitogen-stimulated T cells exhibited chemokinetic behavior toward fibronectin alloyed to Type I collagen, while PMA-stimulated cells did not. The development of fibronectin-receptor activity may be a feature of the more advanced cell cycle progression which occurs prior to the expression of mitogen-induced invasiveness.

This report demonstrates the usefulness of acridine orange cell cycle analysis in elucidating events occurring prior to the onset of S phase. The analysis is especially powerful when it can be supplemented and sharpened by reference to other markers of cell cycle transit, in this case, IL-2R and $gp90^{MEL-14}$. Acridine orange analysis can potentially give better resolution of G_1 subphases than we obtained in these experiments (30). The presence of multiple lymphocyte subsets in our test populations may have blurred the distinctions.

Myriad structural and regulatory genes have been implicated in the promotion and suppression of invasiveness in malignant cells. We hypothesize that at least some of these genes represent altered or inappropriately expressed versions of loci whose original adaptive value was to produce and modulate lymphocytic invasiveness. The findings of this investigation suggest that tests of this hypothesis should center on genes at the onset of G_1, as that is when invasiveness develops.

ACKNOWLEDGEMENTS

This work was supported by grant IM514 from the American Cancer Society. We thank the Michigan Cancer Foundation Flow Cytometry Facility for assistance and analysis.

REFERENCES

1. Hamann A, Jablonski-Westrich D, Scholz K-U, et al: Regulation of lymphocyte homing. I. Alterations in homing receptor expression and organ-specific high endothelial venule binding of lymphocytes upon activation. J. Immunol. 140:737-743, 1988.
2. Jung TM, Gallatin WM, Weissman IL, Dailey MO: Down-regulation of homing receptors after T cell activation. J. Immunol. 141:4110-4117, 1988.
3. Steen PD, McGregor JR, Lehman CM, Samlowski WE: Changes in homing receptor expression on murine lymphokine-activated killer cells during IL-2 exposure. J. Immunol. 143:4324-4330, 1989.
4. Dustin ML, Springer TA: Role of lymphocyte adhesion receptors in transient interactions and cell locomotion. Ann. Rev. Immunol. 9:27-66 1991.
5. Pober JS, Cotran RS: Immunologic interactions of T

lymphocytes with vascular endothelium. Adv. Immunol. 50:261-302, 1991.

6. Shimizu Y, Shaw S: Lymphocyte interactions with extracellular matrix. FASEB J. 5:2292-2299, 1991.

7. Parrott DMV, Wilkinson PC: Lymphocyte locomotion and migration. Progr. Allergy 28:193-284, 1981.

8. Hamann A: Mechanisms of lymphocyte traffic and cell targeting. Int. J. Cancer Suppl. 7:19-23, 1992.

9. Pitzalis C, Kingsley GH, Covelli M, et al: Selective migration of the human helper-inducer memory subset: Confirmation by in vivo cellular kinetic studies. Eur. J. Immunol. 21:369-376, 1991.

10. Cush JJ, Pietschmann P, Oppenheimer-Marks N, Lipsky PE: The intrinsic migratory capacity of memory T cells contributes to their accumulation in rheumatoid synovium. Arthritis Rheum. 35:1434-1444, 1992.

11. Mackay CR: Migration pathways and immunologic memory among T lymphocytes. Semin. Immunol. 4:51-58, 1992.

12. Akbar AN, Amlot PL, Timms A, et al: The development of primed/memory CD8$^+$ lymphocytes in vitro and in rejecting kidneys after transplantation. Clin. Exp. Immunol. 81:225-231, 1990.

13. Mobley JL, Dailey MO: Regulation of adhesion molecule expression by CD8 T cells in vivo. I. Differential regulation of gp90E^{MEL-14} (LECAM-1), Pgp-1, LFA-1, and VLA-4a during the differentiation of cytotoxic T lymphocytes induced by allografts. J. Immunol. 148:2348-2356, 1992.

14. Pilaro AM, Sayers TJ, McCormick KL, et al: An improved in vitro assay to quantitate chemotaxis of rat peripheral large granular blood lymphocytes. J. Immunol. Methods 135:213-224, 1990.

15. Arencibia I, Sundqvist K-G: Collagen receptor in T lymphocytes and the control of lymphocyte motility. Eur. J. Immunol. 19:929-934, 1989.

16. Davis JM, St John J, Cheung HT: Haptotactic activity of fibronectin on lymphocyte migration in vitro. Cell. Immunol. 129:67-69, 1990.

17. Somersalo K, Saksela E: Fibronectin facilitates the migration of human natural killer cells. Eur. J. Immunol. 21:35-42, 1991.

18. Wilkinson PC: Leukocyte locomotion and accumulation: The contributions of cell polarity and cell growth. In: Leukocyte Emigration and its Sequelae. Satellite Symposium of the Sixth International Congress of Immunology, Z Movat (ed), Karger, Basel, pp. 1-13, 1986.

19. Klein V, Kantwerk-Funke G, Zanker KS: Two-dimensional and three-dimensional behavior of native or stimulated peripheral blood lymphocytes. Proc. AACR 31:299, 1990.

20. Applegate KG, Blach CM, Pellis NR: In vitro migra-

tion of lymphocytes through collagen matrix: Arrested locomotion in tumor-infiltrating lymphocytes. Cancer Res. 50:7153-7158, 1990.

21. Ratner S, Patrick P, Bora G: Lymphocyte development of motility in extracellular matrix during IL-2 stimulation. J. Immunol. 149:681-688, 1992.

22. Ratner S: Lymphocyte migration through extracellular matrix. Invasion Metastasis 12:82-100, 1992.

23. Laskin DL, Gardner CR, Laskin JD: Induction of chemotaxis in mouse peritoneal macrophages by activation of protein kinase C. J. Leukocyte Biol. 41:474-480, 1987.

24. Prpic V, Uhing RJ, Weiel JE, et al: Biochemical and functional responses stimulated by platelet-activating factor in murine peritoneal macrophages. J. Cell Biol. 107:363-372, 1988.

25. Blood CH, Zetter BR: Membrane-bound protein kinase C modulates receptor affinity and chemotactic responsiveness of Lewis lung carcinoma sublines to an elastin-derived peptide. J. Biol. Chem. 264:10614-10620, 1989.

26. Schwartz GK, Redwood SM, Ohnuma T, et al: Inhibition of invasion of invasive human bladder carcinoma cells by protein kinase C inhibitor staurosporine. J. Natl. Cancer Inst. 82:1753-1756, 1990.

27. Wilkinson PC, Lackie JM, Haston WS, Islam LN: Effects of phorbol esters on shape and locomotion of human blood lymphocytes. J. Cell Science 90:645-655, 1988.

28. Keller HU, Niggli V, Zimmerman A: Diacylglycerols and PMA induce actin polymerization and distinct shape changes in lymphocytes. Relationship to fluid pinocytosis and locomotion. J. Cell Science 93:457-465, 1989.

29. Oppenheimer-Marks N, Davis LS, Lipsky PE: Human T lymphocyte adhesion to endothelial cells and trans-endothelial migration. Alteration of receptor use relates to activation status of both the T cell and the endothelial cell. J. Immunol. 145:140-148, 1990.

30. Darzynkiewicz Z: Differential staining of DNA and RNA in intact cells and isolated cell nuclei with acridine orange. Methods in Cell Biology 33:285-298, 1990.

31. Gilbert KM, Ernst DN, Hobbs MV, Weigle WO: Effects of tolerance induction on early cell cycle progression by TH1 clones. Cell. Immunol. 141:362-372, 1992.

32. Wilkinson PC: The locomotor capacity of human lymphocytes and its enhancement by cell growth. Immunol. 57:281-289, 1986.

33. Wilkinson PC, Watson EA: FK506 and pertussis toxin distinguish growth-related locomotor activation from attractant-stimulated locomotion in human

blood lymphocytes.   Immunol. 71:417-422, 1990.

34.   Ratner S: Interleukin-2-stimulated lymphocytes.
      Relationship between motility into protein matrix
      and in vivo localization in normal and neoplastic
      tissue.   J. Natl. Cancer Inst. 82:612-615, 1990.

35.   Grzesiak JJ, Davis GE, Kirchhofer D, Pierschbacher
      MD: Regulation of alpha-2 beta-I-mediated fibro-
      blast migration on Type I collagen by shifts in the
      concentrations of extracellular Mg2+ and Ca2+.   J.
      Cell Biol. 117:1109-1117, 1992.

5

FLOW CYTOMETRIC MONITORING OF DRUG RESISTANCE IN HUMAN SOLID TUMORS

Awtar Krishan, Cheppail Ramachandran, Antonieta Sauerteig

Resistance to cancer chemotherapy continues to be a major hurdle in successful management of refractory human malignancies. Drug resistance may be intrinsic or acquired after chemotherapy. Several well-known extra-cellular factors such as drug metabolism and pharmacokinetics may be responsible for failure of chemotherapy. However, a major reason for drug resistance resides at the cellular level and often involves cellular mechanisms which under normal conditions may have other protective and important biological roles. Tumor cell resistance is believed to be multifactorial involving altered drug transport (influx, retention and efflux), and biochemical mechanisms such as xenobiotic detoxification, alternate metabolic pathways, and altered targets (1-3). Multiple drug resistance (MDR) has been recently described as a phenomenon in which tumor cells are resistant to a variety of unrelated natural products such as alkaloids and antibiotics used as cancer chemotherapeutic agents (1,3). Rapid energy dependent drug efflux, the major mechanism involved in MDR, causes reduction in intracellular retention of the chemotherapeutic agent and thereby confers resistance. Several relatively non-toxic drugs have been shown to compete for this elflux pump and thereby enhance drug retention and chemosensitivity.

The attempts to understand the molecular mechanisms

responsible for MDR in human tumor cells have resulted
in the isolation and characterization of two multidrug
resistance genes (MDR1 and MDR3) from human cells (4,5).
Both co-amplification and overexpression of MDR1 and
MDR3 genes in drug resistant tumor cells have been
described (6-9). However, only transfection of the MDR1
gene was able to confer drug resistance in sensitive
cells (10) even though 85% homology of base pairs in the
coding region of the two genes has been reported (10).
The cDNAs of both genes code for P-glycoproteins with
identical structure containing twelve hydrophobic membrane spanning domains organized into two sets of six.

MDR1 gene amplification is frequently seen in cell
lines selected by exposure to increasing drug concentrations in vitro. In human solid tumors, resistance is
usually lower and in cell lines established from refractory tumors with acquired or intrinsic drug resistance,
MDR1 gene amplification is not a common phenomenon. In
our laboratory, both doxorubicin sensitive murine leukemic P388 and 84-fold doxorubicin resistant P388/R-84
cells carry almost equal number of MDR1 gene copies.
Similarly, human melanoma cell lines, such as FCCM-2,
NH, and FCCM-9, established in our laboratory with 2.7-
to 6.1-fold doxorubicin resistance (as compared to the
most sensitive melanoma cell line), do not show any
significant amplification of the MDR1 gene (11).

Overexpression of MDR1 mRNA with or without gene
amplification contributes to drug resistance in solid
tumor cells. In five human melanoma cell lines with 1-
to 6-fold doxorubicin resistance studied in our laboratory, a strong correlation between drug resistance and
MDR1 mRNA content was seen. Two of these cell lines had
55% and 63% P-glycoprotein positive cells, respectively,
when analyzed by flow cytometry after reacting with P-
glycoprotein specific C219 monoclonal antibody. However, P-glycoprotein seemed to be relatively non-function-

al as both of these cell lines showed high doxorubicin retention and efflux blockers did not have a major effect in increasing either the drug retention or cytotoxicity in soft agar colony assays (11).

MDR1 gene encodes a membrane bound 170 Kd P-glycoprotein, which in turn is believed to act as an efflux pump. This efflux pump is versatile in the fact that in its generic form, it seems to handle efflux of a variety of unrelated compounds such as alkaloids, antibiotics, carcinogens, dyes (e.g. Hoechst 33342) and reagents (Indo-AM, Fig. 1). Several in vitro studies have shown that non-toxic concentrations of drugs such as calcium channel blockers (e.g. verapamil), calmodulin inhibitors (phenothiazines), cyclosporin, or dipyridamole will block cellular efflux of a chemotherapeutic agent such as doxorubicin, vinblastine or etoposide and thus enhance retention and chemosensitivity to these drugs (12-16).

Monitoring of cellular drug retention and efflux with analytical methods such as radiolabelled drug uptake, high pressure liquid chromatography and spectrofluorometry is time consuming, require large samples and measure drug retention of the total population. Laser flow cytometry offers a powerful means for monitoring drug retention of fluorescent antitumor drugs such as anthracyclines on a single cell basis (17). This methodology is rapid and identifies heterogeneity in drug retention (Fig. 2) as well as allows sorting of subpopulations of interest for further study.

As most anthracyclines are fluorescent and can be excited with the 488 nm line of an argon ion laser, flow cytometry can be a rapid method for monitoring anthracycline transport (influx, retention and efflux) as well as the effect of various efflux blockers on cellular retention. In earlier studies, we have shown that

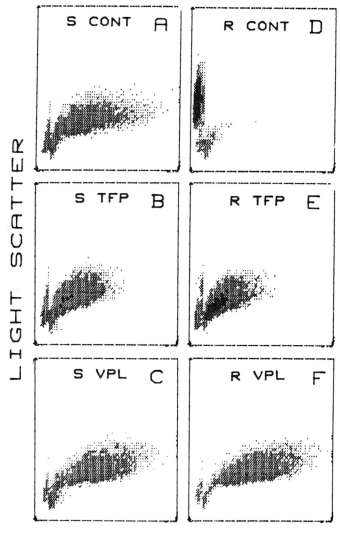

LIGHT SCATTER

INTRACELLULAR FLUORESCENCE

Figure 1. Scattergrams of P388 (A,B,C) and P388/R (D, E,F) cells incubated for 45 min with Indo-AM (A,D), Indo-AM plus 15 μM TfP (B,E), or Indo-AM plus 100 μM VpL (C,F). P388/R-84 cells which have low Indo-AM retention (D) have a significantly higher amount of total fluorescence when incubated in the presence of efflux blockers (E,F).

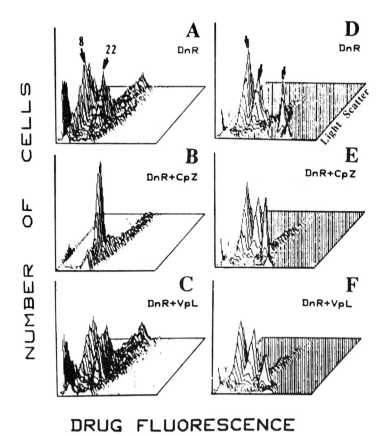

DRUG FLUORESCENCE

Figure 2. In cells recovered from the pretherapy pleural fluid of a lung cancer patient, major heterogeneity in light scatter and DnR fluorescence was seen (A). Incubation with CpZ (B) but not VpL (C) increased drug retention in the subpopulation with the lower retention. In pleural fluid from another lung cancer patient, several subpopulations based on light scatter and DnR retention were seen (D). Modulators CpZ or VpL had no major effect (E,F). (Reproduced with permission from ref. 21).

whereas certain anthracyclines quench their fluorescence on binding to DNA, other anthracyclines such as AD-32 seem to enhance their cellular fluorescence (18). Similarly, the inhibitory effect of anthracyclines on binding of other fluorochromes such as Hoechst 33342 (19) and propidium iodide (20) has been earlier report-

ed. We have earlier reported that in several human
solid tumor cells, extensive heterogeneity in cellular
anthracycline retention (Fig. 2) and sensitivity to
efflux blockers is seen (21). In some human tumor cell
populations, cellular drug retention was very high
compared to that of the control cell line used (Murine
Leukemic P388) whereas in others, cellular drug retention was low. In cells co-incubated with efflux blockers such as phenothiazines or verapamil, cellular retention in selective tumor subpopulations could be significantly increased whereas in other tumors, efflux blockers had no effect. There was also pronounced variation
noted in the response of tumor cells to the efflux
blockers used such as chlorpromazine and verapamil. For
example, in some tumors, verapamil had no effect on drug
retention of subpopulations that were otherwise sensitive to the efflux blocking action of chlorpromazine
(Fig. 2A-C).

While monitoring the effect of various efflux
blockers on cellular retention of doxorubicin, we were
impressed by the fact that phenothiazines such as prochlorperazine would enhance doxorubicin retention in tumor
cells where verapamil had no effect. Based on these
laboratory studies, our clinical colleague, Dr. Kasi
Sridhar, devised a clinical phase I protocol to study
the maximum tolerated dose of prochlorperazine in human
patients. We chose prochlorperazine as it is a potent
doxorubicin efflux blocking antiemetic with low neuro-
toxicity, wide therapeutic index, and high IV dose
tolerance. In our initial phase I protocol, we used
escalating doses of prochlorperazine (15-75 mg/m^2) followed by doxorubicin (60 mg/m^2), both delivered IV over
15 min (22). In a subsequent phase I study, we used
doxorubicin (60 mg/m^2 IV over 15 min.) followed by 75-
180 mg/m^2 of prochlorperazine over 120 mins. We
achieved very high (>1000 ng/ml) plasma levels of proch-

lorperazine in some of our patients. However, there was a great variability in plasma levels achieved and the maximum tolerated dose achieved without any major toxicity was 135 mg/m^2 given over two hours as an IV infusion. We observed activity of this combination in non-small cell carcinoma and mesotheliomas. We have recently activated phase II protocols for modulation of drug resistance in breast cancer, renal cell carcinoma, and other refractory tumors by use of prochlorperazine as an efflux blocker for doxorubicin. Histograms in Fig. 3 show that in some of the tumor specimens retrieved from patients after infusion of prochlorperazine, enhanced doxorubicin retention could be seen (22). We believe this enhanced retention may be due to effective blocking of doxorubicin efflux by prochlorperazine _in vivo_.

CONCLUSION

Our flow cytometric studies on doxorubicin retention and efflux in human solid tumors have shown that extensive heterogeneity exists in cellular retention of this important antitumor agent. Subpopulations in a tumor may react differently to co-incubation with an elflux blocker. Some subpopulations may be sensitive to an efflux blocker while others seem to be unaffected. We have also reported that in tumor subpopulations, rapid modulation of drug retention characteristics and sensitivity to efflux blockers may happen during the course of therapy (22). These studies indicate that laser flow cytometric determination of anthracycline retention and its modulation could be an important tool for selection of efflux blockers which may enhance cellular drug retention and chemosensitivity.

Besides monitoring of drug efflux, laser flow cytometry can be an important tool for detection of cells with P-glycoprotein expression. Recently, several antibodies have become commercially available for immu-

56

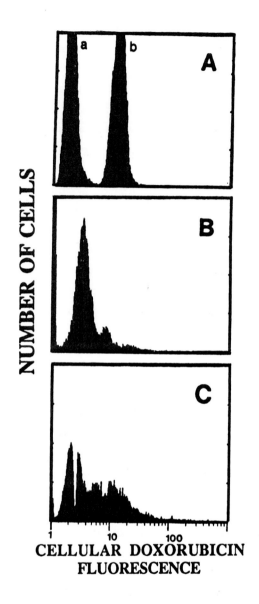

Figure 3. Shows cellular DOX fluorescence of cells from pleural fluid of a patient after Percoll gradient separation. DOX fluorescence in resistant P388/R-84 cells alone (3Aa) or in the presence of an efflux blocker, 25 μM PCZ (3Ab). Cells retrieved from the pleural effusion prior to the start of therapy had DOX retention intermediate between that of P388/R-84 cells with or without incubation with PcZ (3A,B). Sample obtained after 75 min from start of therapy (15 min of PCZ followed by 15 min of DOX) shows presence of cells with enhanced DOX retention (3C). (Reproduced from ref. 22).

nocytochemical and flow cytometric detection and quantitation of P-glycoprotein. We have, in a recent publication (23), compared the flow cytometric utility of three commercially available P-glycoprotein specific antibodies. By using two parameters (DNA content vs P-glycoprotein expression), we can use this method for detection of P-glycoprotein positive aneuploid tumor cells (Fig. 4).

Cellular detoxification of xenobiotics and free radicals also constitute an important mechanism that protects cells from the cytotoxic effects of a drug or other physical (e.g. radiation) agents (24-26). Chemotherapeutic agents such as anthracyclines are believed to cause cellular damage by generation of free radicals. Glutathione (GSH), a tripeptide thiol, plays several important protective roles and biological functions. One of its major roles is the protection of tumor cells against free radicals and electrophiles through both non-enzymatic and enzymatic activities. Elevated levels of GSH and GSH related enzymes such as GSH peroxidase, GSH reductase and glutathione-S-transferase have been reported in several drug resistant cells. The role of these enzymes in the detoxification mechanism is well-established (27). Glutathione peroxidase provides protection from oxidants (26) whereas the glutathione-S-transferase isoenzymes (α, μ and π) detoxify electrophylic xenobiotics by catalization of their conjugation with GSH or through non-catalytic binding of electrophiles (24,27-29).

Laboratory studies have shown that depletion of cellular GSH content by buthionine sulfoxamine (BSO) enhances chemosensitivity (26,30-32). Our investigations have demonstrated that in doxorubicin resistant murine leukemic P388 cells, block of efflux by trifluo perazine reduced cellular resistance from 150- to 60-fold. A combination of GSH depletion (by incubation of

58

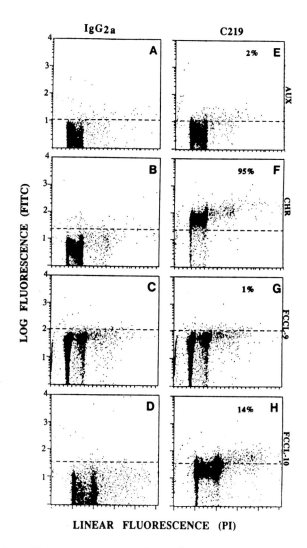

Figure 4. Two parameter dot plots of AUX, CHR, and two human lung cell lines (FCCL-9 and 10) stained by the double staining method for DNA content and P-glycoprotein expression (FACScan). (Reproduced with permission from ref. 23).

cells with BSO for 24 hrs) with efflux blocking by trifluoperazine further reduced resistance from 150- to 8-fold demonstrating the multifactorial nature of cellular resistance to doxorubicin in this model (26).

Several workers have shown that changes in GSH content can be monitored by laser flow cytometry using monochlorobimane (MBCL) as a GSH conjugating fluorochrome (33-35). In most cases, this method gives excellent results and allows for monitoring of GSH content and its heterogeneity in tumor cells. As the GSH-MBCL method is performed on live cells, this method can not only measure changes in GSH content but also be used in a multiparametric setting to analyze heterogeneity as well as correlate GSH content with other phenotypic markers. We have used this technique and established procedures for simultaneous monitoring of doxorubicin retention and GSH content (Fig. 5). As conjugation of MBCL with GSH is dependent on glutathione transferase, one can use time as a parameter to determine the rate for appearance of the fluorescent conjugated MBCL-GSH product. Fig. 6 shows a comparison of the flow cytometric method (time vs GSH-MBCL fluorescence) with the inset showing data obtained from an enzymatic assay method. It is clear that the doxorubicin resistant cells not only had a faster conjugation rate but also the total GSH-MBCL content was higher than that of the sensitive cells.

In conclusion, laser flow cytometry can be a valuable tool for monitoring of drug resistance in human solid tumor cells. Methods are available for determining cellular retention and efflux of a fluorescent drug (e.g. doxorubicin, rhodamine, Hoechst 33342) in resistant cells as well as the effect of various efflux blockers which may enhance cellular drug retention and chemosensitivity. With the help of several commercially available P-glycoprotein specific antibodies, one can identify expression of this efflux pump and by using two-parameter analysis (e.g. DNA content or other phenotypic markers), identify subpopulations for sorting and further in-depth analysis. Glutathione content and

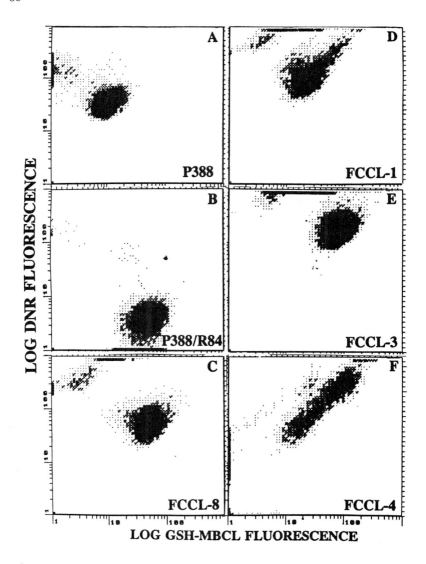

Figure 5. Two parameter dot plots of DNR and GSH-MBCL
fluorescence show low retention of DNR with high GSH-
MBCL fluorescence in P388/R-84 cells (B). In contrast,
all human tumor cell lines examined had higher GSH-MBCL
fluorescence and DNR retention than that of the P388
cells (A). Some of the human tumor cell lines (E,F)
show extensive heterogeneity in DNR retention as well as
in GSH-MBCL fluorescence. (Reproduced with permission
from ref. 35).

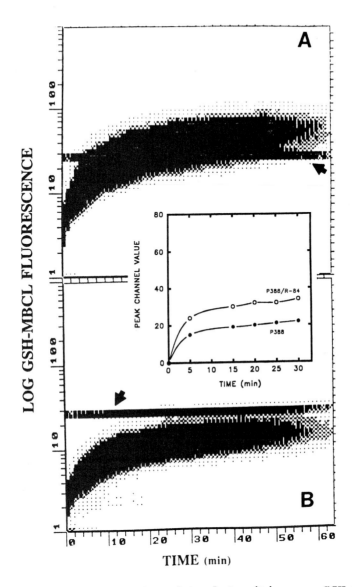

Figure 6. Two parameter dot plots (time vs GSH-MBCL fluorescence) of (A) P388/R-84 and (B) P388 cells. Inset shows peak channel values of GSH-MBCL fluorescence at various time points. Arrows point to fluorospheres used as an internal standard. (Reproduced with permission from ref. 35).

glutathione-S-transferase activity can be analyzed in combination with doxorubicin retention studies to monitor presence of efflux and cellular detoxification, the

two most important mechanisms for drug resistance in
solid tumors.

ACKNOWLEDGEMENT

Supported by NIH grant CA 29360.

REFERENCES

1. Moscow J, Cowan KH: Multidrug resistance. J.
    Natl. Cancer Inst. 80:14-20, 1988.
2. Chabner BA, Fojo A: Multidrug resistance: P-glyco-
    protein and its allies-the elusive foes. J. Natl.
    Cancer Inst. 81:910-913, 1989.
3. Kessel D (ed): Resistance to antineoplastic drugs.
    CRC Press Inc., Boca Raton, 1989.
4. Chen C, Chin JE, Ueda K, et al: Internal duplica-
    tion and homology with bacterial transport proteins
    in the mdr 1 (P-glycoprotein) gene from multidrug-
    resistant human cells. Cell 47:381-389, 1986.
5. Ueda K, Cardarelli C, Gottesman MM, Pastan I:
    Expression of a full-length cDNA for human "MDR1"
    gene confers resistance to colchicine, doxorubicin,
    and vinblastine. Proc. Natl. Acad. Sci. USA, 84:
    3004-3008, 1987.
6. van der Bliek AM, Kooiman PM, Schneioder C, Borst
    P: Sequence of mdr 3 cDNA encoding a human P-gly-
    coprotein. Gene 71:401-411, 1988.
7. Chin J, Soffir R, Noonan K, et al: Structure and
    expression of the human MDR (P-glycoprotein) gene
    family. Mol. Cell Biol. 9:3808-3820, 1989.
8. Raymond M, Rose E, Housman DE, Gros P: Physical
    mapping, amplification and overexpression of the
    mouse mdr gene family in multidrug-resistant cells.
    Mol. Cell Biol. 10:1642-1651, 1990.
9. Shen D, Fojo A, Chin JE, et al: Human multidrug-
    resistant cell lines: Increased mdr-1 expression
    can precede gene amplification. Science 232:643-
    645, 1986.
10. Lincke CR, van der Bliek AM, van der Schuurhuis GJ,
    et al: Multidrug resistance phenotype of human BRO
    melanoma cells transfected with a wild type human
    MDR-1 complementary DNA. Cancer Res. 50:1779-1785,
    1990.
11. Ramachandran C, Yuan ZK, Huang XL, Krishan A:
    Doxorubicin resistance in human melanoma cells:
    MDR-1 and glutathione-transferase p gene expres-
    sion. Biochem. Pharmacol. 45:743-751, 1993.
12. Ganapathi R, Grabowski D: Enhancement of sensitiv-
    ity to Adriamycin in resistant P388 leukemia by the
    calmodulin inhibitor trifluoperazine. Cancer Res.
    43:3696-3699, 1983.

13. Krishan A, Sauerteig A, Wellham L: Flow cytometric studies on modulation of anthracycline transport by phenothiazines. Cancer Res. 45:1046-1051, 1985.

14. Tsuro T: Reversal of acquired resistance to vinca alkaloids and anthracycline antibiotics. Cancer Treat. Rep. 67:889-894, 1986.

15. Slater LM, Sweet P, Stupecki M, et al: Cyclosporin A corrects daunorubicin resistance in Ehrlich ascites carcinoma. Brit. J. Cancer 54:235-238, 1986.

16. Howell SB, Hom D, Sanga R, et al: Comparison of the synergistic potentiation of etoposide, doxorubucin and vinblastine cytoxicity by dipyridamole. Cancer Res. 49:3178-3183, 1989.

17. Krishan A, Ganapathi R: Laser flow cytometric studies on intracellular fluorescence of anthracyclines. Cancer Res. 40:3895-3900, 1980.

18. Krishan A, Ganapathi R: Laser flow cytometry and cancer chemotherapy: Detection of intracellular anthracyclines by flow cytometry. J. Histochem. Cytochem. 27:1655-1656, 1979.

19. Preisler H: Alteration of binding of the supravital dye Hoechst 33342 of human leukemic cells by adriamycin. Cancer Treat. Rep. 62:1393-1396, 1978.

20. Krishan A, Ganapathi R, Israel M: The effect of adriamycin and analogs on the nuclear fluorescence of propidium iodide stained cells. Cancer Res. 38:3656-3662, 1978.

21. Krishan A, Sridhar KS, Davilla E, et al: Patterns of anthracycline retention modulation in human tumor cells. Cytometry 8:306-314, 1987.

22. Sridhar KS, Krishan A, Samy TSA, et al: Prochlorperazine as a doxorubucin-efflux blocker: Phase I clinical and pharmacokinetics studies. Cancer Chemother. Pharmacol. 31:423-430.

23. Krishan A, Sauerteig A, Stein J: A comparison of three commercially available antibodies for flow cytometric monitoring of P-glycoprotein expression in tumor cells. Cytometry 12:731-742, 1991.

24. Meister A, Anderson M: Glutathione. Annual Rev. Biochem. 52:711-760, 1983.

25. Kramer R, Zakher J, Kim G: Role of the glutathione redox cycle in acquired and de novo multidrug resistance. Science 241:694-697, 1988.

26. Nair S, Singh SV, Samy TSA, Krishan A: Anthracycline resistance in murine leukemic P388 cells: Role of drug efflux and glutathione related enzymes. Biochem. Pharmacol. 39:723-728, 1990.

27. Jakoby W, Habig W: In: Enzymatic basis of detoxification, W Jakoby (ed), Academic Press, NY, 2:63-94, 1980.

28. Mannervik A, Alin P, Guthenberg C: Identification of three classes of cytosolic glutathione transferase common to several mammalian species: Correlation between structural data and enzymatic proper-

ties.   Proc. Natl. Acad. Sci. USA   82:7202-7206, 1985.

29.   Litwack G, Ketterer B, Arias I: Ligandin: A hepatic protein which binds steroids, bilirubin, carcinogens and a number of exogenous organic anions. Nature 234:466-467, 1971.

30.   Green J, Vistica D, Young R: Potentiation of mephalan cytotoxicity in human ovarian cancer cell lines by glutathione depletion.   Cancer Res. 44: 5427-5431, 1984.

31.   Russo A, Mitchell JB: Potentiation and protection of doxorubicin cytotoxicity by cellular glutathione modulation.   Cancer Treat. Rep. 69:1293-1296, 1985.

32.   Hromas RA, Andrews PA, Murphy MP, Burns PC: Glutathione depletion reverses cisplatin resistance in murine L1210 leukemia cells.   Cancer Lett. 34:9-13, 1987.

33.   Rice GC, Bump EA, Shrieve DC, et al: Quantitative analysis of cellular glutathione by flow cytometry utilizing monochlorobimane: Some applications to radiation and drug resistance.   Cancer Res. 46: 6105-6110, 1986.

34.   Shrieve DC, Bump EA, Rice GC: Heterogeneity of cellular glutathione among cells derived from a murine fibrosarcoma or a human renal cell carcinoma detected by flow cytometry.   J. Biol. Chem. 263: 14107-14114, 1988.

35.   Nair S, Singh SV, Krishan A: Flow cytometric monitoring of glutathione content and anthracycline retention in tumor cells.   Cytometry 12:336-342, 1991.

6

FLOW CYTOMETRY IN ONCOLOGY: SOME LESSONS FROM HISTORY

Howard M. Shapiro

INTRODUCTION

Although potential applications to cancer diagnosis, research, and treatment provided much of both the
rationale and the funding for the early development of
flow cytometry, a look back through the literature
reveals that the field has not developed precisely as
its pioneers envisioned. While flow cytometry is now
widely used in oncology, reexamining some longstanding
unsolved problems and neglected areas from the vantage
point of the present may be helpful in broadening the
scope of future applications.

CYTODIAGNOSIS: STILL ELUSIVE AFTER ALL THESE YEARS

By the 1950's, Caspersson (1) and Papanicolaou (2)
and their coworkers had established the existence and
clinical relevance of differences in nucleic acid content between normal and malignant cells, setting the
stage for the development of a series of static microph-
otometric instruments (3,4) intended to make diagnostic
cervical cytology more objective and automatic.

Kamentsky's development of the first quantitative
multiparameter optical flow cytometers in the 1960's was
a direct result of his previous experience with image
analysis; it was clear to him that the technology of the
time precluded using high-resolution cell scanning and
feature extraction by image processing as the basis for

a clinical laboratory instrument. Since nucleic acid
content and cell size had been established as useful
parameters for cervical cell classification, Kamentsky's
original flow cytometer (5) was designed to measure the
first of these parameters by absorption at 260 nm and
the second by light scattering at 410 nm. A clinical
trial of this instrument on cervical cytology specimens
was done (6), and showed promising results; another
group reported similar results using a similar instrument (7).

By the late 1960's, flow cytometric measurements of
DNA content using fluorescent reagents had been proposed
by Sandritter (8) and demonstrated by Dittrich and Gohde
(9) and by Van Dilla et al (10); fluorescence measurement capability was incorporated into instruments produced by Kamentsky's firm, Bio/Physics Systems (later
acquired by Ortho) and by Phywe, which commercialized
the Dittrich/Gohde instrument. However, it seemed clear
at that time that acceptable performance in cervical
cancer cytodiagnosis or screening would require more
than the two- or three-parameter measurement capabilities of the first commercial flow cytometers.

Since work on image analysis of cervical cytology
specimens had shown that accuracy improved as increasing
numbers of morphologic parameters were included in cell
classification algorithms, it was felt that the addition
of spatial resolution to flow cytometers would improve
their performance. Several approaches were taken during
the 1970's; Wheeless and his collaborators in Rochester
built increasingly elaborate slit-scanning instruments
allowing multidimensional imaging of cells in flow (11),
while the Los Alamos group examined time-of-flight (12)
and multiangle scatter measurements (13). None of these
approaches yielded an apparatus which could be seriously
considered for clinical use, nor did the addition by
Darzynkiewicz et al of RNA content measurement to DNA

content measurement (14).

By the mid-1980's, interest in flow cytometric automation of cervical cancer cytodiagnosis had waned considerably, even as techniques originally developed to attack the problem found increasingly wide application in related areas of oncology. Determination of DNA content abnormalities in a variety of solid tumors became accepted clinical practice (15,16), and DNA content measurements which had been tried and found wanting for bladder cancer screening proved useful for monitoring therapy (17). As apparatus for computer image analysis simultaneously improved in speed and quality and dropped in price, work on the development of automated cervical cytology systems shifted toward apparatus to improve preparation and interpretation of the conventional Papanicolaou smear.

From the vantage point of 1993, we might well ask whether this makes sense. The flow cytometry of the 1970's, lacking capacity for analyses of more than two or three parameters, and without today's monoclonal antibodies and gene probes, was clearly not up to the task of cervical cytodiagnosis, but it is not clear that the use of neural networks or other advanced software methods to interpret data from slides stained with non-specific dye mixtures developed in the 1940's will produce drastic improvements over what cytotechnologists and pathologists can now achieve. Whether used with flow cytometry or image analysis, it seems likely that a combination of reagents which measured nucleic acid content, virus- and/or proliferation-associated anti-gens, and oncogenes or their products could provide more accurate diagnostic information.

Why, then, has no work been done along this line? In all probability, because the National Cancer Insti-tute, the principal source of funding for earlier ef-forts at flow cytometric cytodiagnosis, learned a lesson from history in the 1980's which may not be valid today. With grant funding essentially unavailable and the concept not sufficiently well defined to generate investment by the private sector, further progress is unlikely, but not impossible.

DRUG EFFECTS ON CANCER CELLS: A MOVING TARGET

Although only a few years elapsed between the publication of the first flow cytometric DNA content histograms and the description of the effects of various anticancer agents on the cell cycle (18,19), the problem of determination of drug effects in clinical situations remains only partially solved.

It was initially hoped that the cytometric definition of differences in cell cycle kinetics between tumor cells and normal stem cell populations would allow the development of treatment schedules with improved therapeutic indices. However, although flow cytometry produced impressive refinements in methods of cytokinetic analysis, including increasingly sophisticated mathematical modeling to determine S-phase fraction (20), definition of subcompartments of the cell cycle based on DNA and RNA content and chromatin structure (21), and bromodeoxyuridine labeling techniques equal to or better than tritiated thymidine labeling for detection of DNA synthesis (22,23), the kinetics of tumor cells thereby revealed did not provide the hoped-for basis for drug schedule design.

Attempts were also made to characterize stem cell populations within tumors in order to develop drug sensitivity tests analogous to those now used clinically to determine antibiotic susceptibilities of microorganisms. Two principal problems arose in this area. One was reducing solid tumor samples to the monodisperse cell suspensions desirable for flow cytometric analysis while simultaneously preserving the cell viability

required for subsequent analysis of drug effects in
culture and allowing vital staining with enough reagents
to delineate stem cells. The second was the definition
of cytometric indicators of cell injury and death suitable for use in the context of anticancer drug treatments, which typically compromise reproductive viability
long before other metabolic and morphologic evidence of
damage becomes apparent. Both problems remain to some
extent, although the direction of research on drug
interactions with cancer cells has shifted.

Sensitivity assays, cytometric and otherwise, have
failed to attain widespread clinical use because most
cancers are insensitive to most drugs. Flow cytometry
has, however, helped to define some mechanisms of drug
resistance. Fluorescent analogues of methotrexate were
used to study the gene amplification involved in resistance to this agent (24,25), and the natural fluorescence of the important anthracycline family of anticancer agents led Krishan (26) and subsequently others to
use flow cytometry to monitor cells' uptake of these
agents. There is now widespread interest in the glycoprotein efflux pump responsible for resistance to anthracyclines and many other drugs, and such multidrug
resistance can now be detected and quantified by measurement of cells' uptake and loss of a variety of
fluorescent compounds (27,28,29) as well as by the use
of fluorescently labeled antibodies to the glycoprotein
itself (30). Cellular glutathione content measurements
(31) have also been used as indicators of drug resistance.

Among the dyes now used for demonstration of the
efflux pump are rhodamine 123 and the cyanine dyes,
previously described (32) as indicators of cytoplasmic
and mitochondrial membrane potential and, as such, used
to detect putative susceptibility of cancer cells to
drugs (33) and early indications of cytotoxic effects

(34). Some of the earlier work done with these dyes should be reexamined to determine whether fluorescence changes attributed to changes in membrane potential might have been due to changes in the activity of the efflux pump and vice versa.

So-called "viability" tests based on membrane integrity as evidenced by the capacity of intact cells to exclude dyes such as trypan blue and propidium iodide are readily adapted to flow cytometry, but are recognized as not well suited for the determination of drug effects on cancer cells (35). In recent years, interest in programmed cell death or apoptosis has led to the development of a variety of flow cytometric methods for detection of nucleic acid degradation and other changes associated with this phenomenon (36).

Although flow cytometric assays for drug resistance and cell death have yet to assume significant clinical roles, work in this area is very much in the mainstream, making it fundable and providing a basis for optimism.

CYTOMETRIC SCREENING FOR CANCER: THE STRANGE STORY OF THE CERCEK TEST

There are few concepts which diagnostics companies find as attractive as the prospect of a generalized test which could detect a pea-sized cancer anywhere in the body. The Cercek or SCM test for cancer, a cytometric test originally conceived for this purpose, has had adherents and detractors, but has attracted commercial support from numerous sources since the mid-1970's. SCM is an abbreviation for "structuredness of cytoplasmic matrix"; the term was coined by Lea and Boris Cercek (37,38) to describe a functional cellular parameter which is quantified by measuring the polarization of fluorescence of intracellular fluorescein following incubation of cells with fluorescein diacetate (FDA).

FDA, which is actually the diacetyl ester of fluorescein, has been widely used in tests for cell "viability", i.e., membrane integrity; the ester, which is uncharged, lipophilic, and nonfluorescent, readily crosses intact cell membranes and is hydrolyzed to the fluorescent fluorescein anion by nonspecific intracellular esterases. Although leakage occurs over a period of minutes to hours, the fluorescein anion is retained by cells with intact cytoplasmic membranes, imparting green fluorescence to the cytoplasm; fluorescein does not accumulate to any significant extent in cells with damaged membranes, which therefore exhibit only minimal fluorescence.

In order for fluorescence to occur, the fluorescent molecule must first absorb light. In order for absorption to occur, the dipole moment of the absorbing molecule must be aligned with the electric field vector of the incident light. If this light is linearly polarized, only those molecules which happen to be oriented properly with respect to the plane of polarization can absorb light, and only those molecules are capable of fluorescence emission. Absorption occurs so rapidly (10^{-15} s) that the absorbing molecules have no time to move during the process. If fluorescence emission occurred as rapidly as absorption, or if the molecules involved were completely immobilized, the fluorescence emission occurring following excitation by linearly polarized light would be linearly polarized. However, fluorescence emission typically does not occur for tens or hundreds of nanoseconds after absorption, and a change in the orientation of an excited molecule during this fluorescence lifetime results in a change cf the direction of polarization of the emitted light. The relative intensities of fluorescence measured in planes of polarization parallel and perpendicular to the plane of polarization of the excitation beam thus provide an indication of the mobility of fluorescent molecules.

When intracellular esterase acts on FDA, some of
the resulting fluorescein anions bind to basic groups on
proteins; the mobility of protein-bound fluorescein
molecules is restricted compared to the mobility of
fluorescein molecules free in solution. Similarly, the
mobility of fluorescein molecules bound to large aggregates of protein molecules is less than the mobility of
fluorescein molecules bound to single protein molecules
or small aggregates. The degree of polarization of
fluorescence from intracellular fluorescein will therefore depend on the ratio of free to protein-bound fluorescein and on the state of aggregation of intracellular
proteins.

The Cerceks measured fluorescein fluorescence
polarization in cell suspensions in cuvettes using a
modified spectrofluorometer, and, in analyses of syn-
chronized cell cultures, found changes in polarization
during different phases of the cell cycle. They then
used conventional density sedimentation techniques to
isolate lymphocytes from peripheral blood, and found
that decreases in fluorescence polarization occurred
within an hour following exposure of these cells to
polyclonal mitogens such as phytohemagglutinin (PHA).
They also noted that lymphocytes from cancer patients
exhibited smaller changes in polarization after exposure
to PHA than did cells from controls, and that lymphocytes from patients with cancer showed polarization
changes after exposure to antigens extracted from tumors, while cells from controls showed little if any
effect following exposure to tumor extracts. These
results seemed to promise a rapid method for cancer
diagnosis, and therefore excited considerable interest.

From about 1975 on, several groups of investigators
attempted to duplicate the Cerceks' results with cancer
patients, with variable results. It seemed logical that
the precision and sensitivity of analyses would be

increased by making measurements of single cells; this has been done by static microfluorometry (39, M. Deutsch, personal communication) as well as by flow cytometry. Stewart et al (40) reported differences in the fluorescein polarization responses of lymphocytes from breast cancer patients and controls following exposure to PHA and pooled tumor antigen; their study was designed to exclude observer bias.

On another front, Miller et al (41) demonstrated changes in fluorescence polarization of bone marrow cells within a few hours following addition of preparations of (granulocyte) colony stimulating factor and erythropoietin. By sorting cells with the most marked responses, they were able to obtain suspensions enriched in either granulocytic or erythrocytic precursors dependent upon which cell growth factor was used as the stimulus. This suggested that changes in fluorescein fluorescence polarization, whatever their physicochemical explanation might be, could serve as general indicators of the effect of activators such as mitogens and growth factors on responsive cells.

Udkoff et al (42) analyzed the effects of the concentration of lectin, FDA, Ca^{++}, and K^+ on the polarization responses of lymphocytes from normal donors, and found that all of the above might influence the shape of both fluorescence polarization and intensity distributions; lymphocytes (in retrospect, probably B-cells) from a patient with chronic lymphocytic leukemia showed no response to PHA. Most recently, Nairn and his co-workers (43,44) demonstrated that the responding cells are T-cells, that human T-cells respond to the mitogenic OKT3 antibody as well as to PHA, and that calcium and an intact cytoskeleton are required for polarization changes to occur.

Although several groups have demonstrated differences in fluorescence polarization responses between cancer patients and individuals without cancer, the Cercek test has not been shown to be reliable enough to be used as a generalized screening test. This has almost inevitably led the companies funding development to stop work. As a result, some interesting findings from studies done to date have been overlooked. Although the phenomenon is not always demonstrable, it appears that, in many cancer patients, a substantial fraction of peripheral T-cells appear to be stimulated or activated following exposure to an antigen or antigens present in tumor cells; the response may reflect either expansion of a clone or clones of tumor-responsive cells or a recruitment phenomenon such as might result from cytokine secretion by a smaller, specifically stimulated subpopulation.

Unfortunately, fluorescence polarization measurements are among the least conventional and most difficult methods for the demonstration of lymphocyte activation, and, as it happens, no one has yet bothered to attempt to correlate early fluorescence polarization responses with more generally accepted indicators of lymphocyte activation such as calcium uptake, activation antigen display, or changes in RNA content, cell size, or cytoplasmic or mitochondrial membrane potential (45). If the Cerceks' original work had been done with calcium probes the 1980's instead of fluorescence polarization in the 1970's, the literature might provide answers instead of provoking questions about the existence, nature, and function of tumor-reactive T-cells in peripheral blood. As it now stands, having not learned from history, we will have to repeat it, but we stand to gain information which may be useful for therapy as well as for diagnosis.

NORMAL AND ABNORMAL HEMATOPOIESIS: LESSONS FOR LAB AND CLINIC

Flow cytometric characterization of normal and abnormal hematopoietic and lymphoid cells encompasses some of the best developed and most elegant flow cytometric techniques now in use. The humble origins of leukemia and lymphoma phenotyping date back to the early 1970's, when polyclonal antibodies, used for single-parameter immunofluorescence measurements, were found to define prognostic categories of acute lymphocytic leukemia in children.

At that time, before monoclonal antibodies had been described and with even two-color immunofluorescence flow cytometry not feasible due to a lack of suitable labels, immunologists had little if any appreciation of the advantages which might be gained by multiparameter flow cytometric analysis of leukemias. The earliest multiparameter work was done by groups involved in the development of flow cytometric differential leukocyte counters (46). A combination of light scattering and fluorescence measurements showed that leukemic cells appeared where few, if any, normal cells were to be found.

During the 1980's and 1990's, multicolor immunofluorescence labeling techniques and multiparameter flow cytometry have defined normal hematopoiesis in terms of patterns of cellular antigen display which are remarkably uniform from individual to individual (47), and demonstrated (47,48) the uniqueness of leukemic clones. Lessons learned from work in this area have facilitated the isolation of normal stem cells (49) for potential therapeutic applications and enabled the detection of minimal residual disease in patients with acute myeloid leukemia in clinical remission.

There may be a less obvious lesson to be learned from further inquiries into the basis of clonal individ-

uality in leukemias. Investigators (M. Loken, C. Stewart, L. Terstappen, personal communications) who have
looked at several hundred leukemic clones from patients
with myeloid leukemia report that no two are exactly
alike phenotypically; this may be of fundamental biological significance. Like other malignant cells, leukemic
cells differ genotypically from their host's normal
cells, and it is generally accepted that several muta-
tional events are necessary for malignant transformation, and that many different combinations of somatic
mutations can give rise to malignancy even in a single
cell type. It is therefore reasonable to expect that
there would be dozens or even hundreds of potential
leukemic genotypes.

At first glance, it does not seem unreasonable
that, if there were several hundred possible leukemic
genotypes, examination of several hundred leukemic
clones would show no two were phenotypically identical.
However, this represents one of many instances in which
conclusions which can easily be reached using probability theory are not intuitively obvious. The leukemic
genotype/leukemic phenotype problem is similar to the
problem of determining how many people one needs to have
at a party before the odds become even that two of those
present have the same birthday. There are 366 possible
birthdays; calculations reveal that the probability is
greater than 0.5 that two people have the same birthday
when there are 24 or more people at the party.

Taking this line of reasoning back to leukemias,
there are two possibilities. The first is that there
are many more than several hundred possible leukemic
genotypes, each matching a phenotype. The second, which
seems more interesting, is that a selection process,
slightly different for each host and presumably mediated
by host factors, is responsible for the emergence of the
predominant leukemic phenotype or phenotypes. While

some leukemic cells proliferate readily in vitro, with such proliferative capacity associated with a poor prognosis (50), it is surprisingly difficult to culture cells from many malignancies, even though the tumors may be highly aggressive.

After more than a quarter of a century of using flow cytometry to look at cancer cells, we still haven't come up with a cytometric definition of what differentiates cancer cells from normal cells, one which holds as well for Pap smears as for marrow aspirates. The information we are able to obtain frequently allows us to prognosticate, but only rarely permits any therapeutic intervention which might substantially alter the predicted outcome. It is possible that we've been focusing our thoughts and apparatus on the malignant cells and overlooking host factors, cellular and otherwise, which are critical to the understanding and control of cancer. We can at least hope to learn some more valuable lessons in the next twenty-five years.

REFERENCES

1. Caspersson TO: Cell Growth and Cell Function. Norton, New York, 1950.
2. Papanicolaou GN, Traut HF: The diagnostic value of vaginal smears in carcinoma of the uterus. Am. J. Obst. Gynec. 42:193, 1941.
3. Mellors RC, Silver R: A microfluorometric scanner for the differential detection of cells: Application to exfoliative cytology. Science 114:356, 1951.
4. Miner RW, Kopac MJ (eds): Cancer Cytology and Cytochemistry. Ann. N.Y. Acad. Sci., 63, 1956.
5. Kamentsky LA, Melamed MR, Derman H: Spectrophotometer: New instrument for ultrarapid cell analysis. Science 150:630, 1965.
6. Koenig SH, Brown RD, Kamentsky LA, et al: Efficacy of a rapid cell spectrophotometer in screening for cervical cancer. Cancer 21:1019, 1968.
7. Finkel GC, Grand S, Ehrlich MP, DeCote R: Cytologic screening automated by Cytoscreener. J. Assn. Adv. Med. Instrum. 4:106-110, 1970.
8. Evans DMD (ed): Cytology Automation. Livingstone, Edinburgh, 1970.

9. Dittrich W, Gohde W: Impulsfluorometrie bei einze-
    lzellen in suspensionen. Z. Naturforsch. 24b:360,
    1969.
10. Van Dilla MA, Trujillo TT, Mullaney PF, Coulter JR:
    Cell microfluorometry: A method for rapid fluores-
    cence measurement. Science 163:1213, 1969.
11. Wheeless LL Jr: Slit-scanning. In: Flow Cytometry
    and Sorting, MR Melamed, T Lindmo, ML Mendelsohn
    (eds), Wiley-Liss, New York, pp. 103-125, 1990.
12. Steinkamp JA, Crissman HA: Automated analysis of
    deoxyribonucleic acid, protein, and nuclear-to-
    cytoplasmic relationships in tumor cells and gynoc-
    ologic specimens. J. Histochem. Cytochem. 22:616-
    621, 1974.
13. Steinkamp JA, Hansen KM, Wilson JS, Salzman GC:
    Automated analysis and separation of cells from the
    respiratory tract: Preliminary characterization
    studies in hamsters. J. Histochem. Cytochem. 25:
    892-898, 1977.
14. Traganos F, Darzynkiewicz Z, Sharpless T, Melamed
    MR: Simultaneous staining of ribonucleic and de-
    oxyribonucleic acids in unfixed cells using acri-
    dine orange in a flow cytofluorometric system. J.
    Histochem. Cytochem. 25:46, 1977.
15. Barlogie B, Drewinko B, Schumann J, et al: Cellu-
    lar DNA content as a marker of neoplasia in man.
    Amer. J. Med. 69:195, 1980.
16. Hedley DW, Friedlander ML, Taylor IW, et al: Meth-
    od for analysis of cellular DNA content of paraf-
    fin-embedded pathological material using flow cyto-
    metry. J. Histochem. Cytochem. 31:1333, 1983.
17. Klein FA, Herr HW, Whitmore WF Jr, et al: Automat-
    ed flow cytometry to monitor intravesical BCG ther-
    apy of superficial bladder cancer. Urology 17:310-
    314, 1981.
18. Tobey RA, Crissman HA: Use of flow microfluorome-
    try in detailed analysis of effects of chemical
    agents on cell cycle progression. Cancer Res.
    32:2726, 1972.
19. Andreeff M (ed): Impulscytophotometrie. Springer,
    Berlin, 1975.
20. Gray JW, Dolbeare F, Pallavicini MG: Quantitative
    cell-cycle analysis. In: Flow Cytometry and Sort-
    ing, MR Melamed, T Lindmo, ML Mendelsohn (eds),
    Wiley-Liss, New York, pp. 445-467, 1990.
21. Darzynkiewicz Z, Traganos F, Melamed MR: New cell
    cycle compartments identified by multiparameter
    flow cytometry. Cytometry 1:98, 1980.
22. Gray JW, Mayall BH (eds): Monoclonal Antibodies
    Against Bromodeoxyuridine, Alan R. Liss, New York,
    1985 (Also published as Cytometry Volume 6, Number
    6).
23. Crissman HA, Steinkamp JA: A new method for rapid
    and sensitive detection of bromodeoxyuridine in DNA

replicating cells. Exp. Cell Res. 173:256, 1987.

24. Mariani BD, Slate DL, Schimke RT: S phase specific synthesis of dihydrofolate reductase in Chinese hamster ovary cells. Proc. Natl. Acad. Sci. USA 78:4985, 1981.

25. Rosowsky A, Wright J, Shapiro H, et al: A new fluorescent dihydrofolate reductase probe for studies of methotrexate resistance. J. Biol. Chem. 257:14162, 1982.

26. Krishan A, Ganapathi R: Laser flow cytometric studies on the intracellular fluorescence of anthracyclines. Cancer Res. 40:3895, 1980.

27. Lalande M, Ling V, Miller RG: Hoechst 33342 dye uptake as a probe of membrane permeability changes in mammalian cells. Proc. Natl. Acad. Sci. USA 78:363-367, 1981.

28. Lampidis TJ, Munck J-N, Krishan A, Tapiero H: Reversal of resistance to rhodamine 123 in adriamycin-resistant Friend leukemia cells. Cancer Res. 45:2626-2631, 1985.

29. Kessel D, Beck WT, Kukuraga D, Schulz V: Characterization of multidrug resistance by fluorescent dyes. Cancer Res. 51:4665-4670, 1991.

30. Krishan A, Sauerteig A, Stein JH: Comparison of three commercially available antibodies for flow cytometric monitoring of P-glycoprotein expression in tumor cells. Cytometry 12:731-742, 1991.

31. Hedley DW: Flow cytometric assays of anticancer drug resistance. Ann. N.Y. Acad. Sci. 677:340-353, 1993.

32. Shapiro HM: Cell membrane potential analysis. Methods Cell. Biol. 33:25-35, 1990.

33. Lampidis TJ, Bernal SD, Summerhayes IC, Chen LB: Selective toxicity of rhodamine 123 in carcinoma cells in vitro. Cancer Res. 43:716, 1983.

34. Bernal SD, Shapiro HM, Chen LB: Monitoring the effect of anti-cancer drugs on L1210 cells by a mitochondrial probe, rhodamine-123. Int. J. Cancer 30:219, 1982.

35. Roper P, Drewinko B: Comparison of in vitro methods to determine drug-induced cell lethality. Cancer Res. 36:2182-2188, 1976.

36. Darzynkiewicz Z, Bruno S, Del Bino G, et al: Features of apoptotic cells measured by flow cytometry. Cytometry 13:795-808, 1992.

37. Cercek L, Cercek B, Ockey CH: Structuredness of the cytoplasmic matrix and Michaelis-Menten constants for the hydrolysis of FDA during the cell cycle in Chinese hamster ovary cells. Biophysik 10:187, 1973.

38. Cercek L, Cercek B: Application of the phenomenon of changes in the structuredness of cytoplasmic matrix (SCM) in the diagnosis of malignant disorders: A review. Eur. J. Cancer 13:903, 1977.

39. Cercek L, Cercek B: Changes in the structuredness of cytoplasmic matrix (SCM) in human lymphocytes induced by phytohaemagglutinin and cancer basic protein as measured on single cells. Brit. J. Cancer 33:359, 1976.

40. Stewart S, Pritchard KI, Meakin JW, Price GB: A flow system adaptation of the SCM test for detection of lymphocyte response in patients with recurrent breast cancer. Clin. Immunol. Immunopathol. 13:171, 1979.

41. Miller RG, Lalande ME, McCutcheon MJ, et al: Usage of the flow cytometer-cell sorter. J. Immunol. Methods 47:13, 1981.

42. Udkoff R, Chan S, Norman A: Identification of mitogen responding lymphocytes by fluorescence polarization. Cytometry 1:265, 1981.

43 Rolland JM, Dimitropoulos K, Bishop A, et al: Fluorescence polarization assay by flow cytometry. J. Immunol. Methods 76:1-10, 1985.

44. Dimitropoulos K, Rolland JM, Nairn RC: Analysis of early lymphocyte activation events by fluorescence polarization flow cytometry. Immunol. Cell Biol. 66:253-260, 1988.

45. Shapiro HM: Practical Flow Cytometry. 3rd Ed., Wiley-Liss (eds), New York, pp. 326-348 and 395-402, 1994.

46. Shapiro HM, Young RE, Webb RH, Wiernik PH: Multiparameter flow cytometric characterizations of cell populations in acute leukemia. Blood 50 Supp. 1: 209, 1977.

47. Terstappen LW, Loken MR: Myeloid cell differentiation in normal bone marrow and acute myeloid leukemia assessed by multi-dimensional flow cytometry. Anal. Cell. Pathol. 2:229-240, 1990.

48. Terstappen LW, Safford M, Konemann S, et al: Flow cytometric characterization of acute myeloid leukemia. Part II. Phenotypic heterogeneity at diagnosis. Leukemia 6:70-80, 1992.

49. Huang S, Terstappen LW: Formation of haematopoietic microenvironment and haemopoietic stem cells from single human bone marrow stem cells. Nature 360:745-749, 1992.

50. Lowenberg B, Van Putten WLJ, Touw IP, et al: Autonomous proliferation of leukemic cells in vitro as a determinant of prognosis in adult acute myeloid leukemia. N. Engl. J. Med. 328:614-619, 1993.

7

DNA FLOW CYTOMETRY APPLICATION TO CLINICAL TRIALS IN
BREAST CANCER

Lynn G. Dressler

INTRODUCTION

The appropriate management of the breast cancer
patient with early stage disease is a controversial,
frustrating issue. Prognostic factors that could accurately predict tumor behavior would greatly aid both the
clinician and the patient in their treatment decisions.
The clinical trial setting affords one of the best
opportunities to assess new treatment strategies as well
as evaluate new prognostic markers. Several prognostic
factors are currently being evaluated in a number of
clinical trials throughout the United States, the most
well established of these markers is DNA flow cytometry.

There are at least two measurements of tumor aggressiveness that we can obtain using DNA flow cytometry
techniques: One is the estimate of the ploidy status or
DNA content of a patient's tumor, usually referred to as
"DNA diploid" if the tumor contains a normal amount of
DNA or "DNA aneuploid" if the tumor contains an abnormal
amount of DNA (hypo or hyperdiploid). The other prognostic factor is an estimate of a tumor's proliferative
capacity, obtained by measuring the percent of nuclei in
the DNA synthetic phase (S phase) of the cell cycle.
The S phase is usually referred to as "high" or "low"
(and sometimes intermediate) and is often based on a
mean or median value obtained from a large database.

DNA flow cytometry measurements of ploidy and

proliferative capacity have had significant clinical impact in predicting relapse-free survival (RFS) in early and late stage breast cancer (1-13). As these studies mature and with the use of paraffin embedded material for retrospective studies, we are also observing their importance in predicting overall survival as well (2,4,5,6,13). Most recently, we have begun to evaluate and observe that these measurements may be useful in predicting response to therapy (14-19).

This chapter will discuss the clinical trials that have incorporated DNA flow cytometry measurements as part of their protocol and will summarize work published relevant to node negative disease.

DNA FLOW CYTOMETRY IN THE NODE NEGATIVE BREAST CANCER PATIENT

One of the goals of our laboratory and many others is to identify breast cancer patients at high risk and at low risk for developing a tumor recurrence. The objective is to enable the most appropriate treatment to be given at the earliest possible time. The management of the breast cancer patient with node negative disease has been a difficult issue, since approximately 70% of these women will be cured by surgery alone. Therefore, it is necessary to try to better identify those women in the remaining 30% who are at risk for developing a recurrence.

There are several prognostic factors that contribute to a patient's risk profile. These factors include: nodal status, tumor size, histopathology, hormone receptor status, patient age and menopausal status. DNA flow cytometry measurements of ploidy and proliferative activity have also been evaluated by several investigators for the ability to predict prognosis in the node negative patient.

Table 1 summarizes the results of eight studies

Table 1

Ploidy And S Phase As Predictors Of RFS And OS In The
Node Negative Breast Cancer Patient
Univariate Analysis

| | | | p Value | | | |
| | | | RFS | | OS | |
REFERENCE	# OF PTS	FU (YRS)	PLOIDY	SPF	PLOIDY	SPF
Winchester (1990)	198	6.67	NS	.03	NS	NS
Muss (1989)	101	4.25	.23	.05	.925	.006
Fisher (1990)	243	10.00	.7	.003	.9	.001
Keyhani-Rofagha (1990)	165	8.00	-	-	.3	-
O'Reilly (1990)	169	8.00	.6	.006	-	-
Clark (1991)	345	6.00	.02	.031 (D) .01 (A)	-	-
Siguurdson (1990)	286	4.00	.03	.007	.005	.005
Lewis (1990)	155	10.00	.0001	-	.0001	-

D=diploid; A=aneuploid

that have measured ploidy and S phase as prognostic
factors in the node negative breast cancer patient.
These studies range in patient number from 100 to nearly
350, with 4-10 years of median follow-up time. Let us
first focus on the ability of ploidy status to predict
RFS. Three of the studies report that ploidy status can
significantly discriminate for RFS, i.e., patients with
diploid tumors have a longer RFS compared to patients
with aneuploid tumors [Clark (7), Siguurdson (5), Lewis
(20)]. The other four studies measuring ploidy status
in relation to RFS, however, show conflicting results:
None of these studies found that ploidy status could
significantly predict for RFS [O'Reilly (10), Winchester
(8), Muss (6), Fisher (21)].

   If we now focus on the column labeled S phase

fraction (SPF) evaluating the ability of SPF to predict
for RFS, we observe that all studies consistently show
SPF to be a significant predictor of RFS.

Six of the studies listed also reported results for
the ability of these measurements to predict overall
survival (OS). Two studies [Siguurdson (5), Lewis (20)]
show a significant relationship for ploidy status to
predict OS, while the other four studies found no significant relationship. Of the four studies measuring
SPF in relation to overall survival, three groups found
SPF to be a significant marker of OS (5,6,21).

Perhaps a more relevant way of summarizing the data
on proliferative capacity measurements is to look at the
relationship in terms of percentage of patients remaining free of disease at five years. Table 2 summarizes
the estimated 5 year disease-free survival for low S-
phase tumors compared to high S-phase tumors. We observe a striking difference in the percent of patients
remaining free of disease between the low and high S-
phase groups. An average difference of 21% was reported
among all the studies. In Winchester's study, S phase

Table 2

Five Year % Disease Free Survival By S Phase[a]

REFERENCE	LOW S		HIGH S		DIFFERENCE	
O'Reilly, 1990	77%		54%		23%	
Siguurdson, 1990[b]	89%		69%		20%	
Muss, 1989	82%		65%		17%	
Winchester, 1990	92%	(D)	73%	(D)	19%	(D)
Clark, 1991	93%	(D)	73%	(D)	20%	(D)
	81%	(A)	64%	(A)	17%	(A)
(Fisher, 1991[c]	46%		27%		19%)	

[a]Estimated from actuarial DFS curves
[b]Estimated at 4 year follow-up
[c]Ten year % DFS obtained from text
D=diploid; A=aneuploid

predicted disease-free survival only in the diploid tumors (8). Clark's update shows that low S phase in both diploid and aneuploid tumors correspond to a significant disease-free survival advantage compared to the high S-phase group (7). Data from the National Surgical Adjuvant Breast and Bowel Project (NSABP) trial shows 10 year not 5 year disease-free survival of 46% for low S phase and 27% for high S phase patients in a group of clinically node negative patients (21). Information was not provided for estimating 5 year disease-free survival in this study.

In summary, those studies measuring ploidy and proliferative activity in the node negative breast cancer patient have shown that: 1) Measurements of proliferative capacity using DNA flow cytometry consistently show SPF to be a significant predictor of disease-free survival; 2) a substantial difference is observed in the percentage of patients remaining free of disease between the low and high S-phase groups; and, 3) measurements of ploidy status show inconsistent results in predicting both RFS, and when measured, OS.

CLINICAL TRIALS USING DNA CYTOMETRY MEASUREMENTS

There are several clinical trials which have or are currently using DNA flow cytometry measurements as part of a clinical or ancillary laboratory protocol. Objectives of these studies include evaluating ploidy and S phase as markers of patient prognosis and in addition evaluating their usefulness to predict response to chemotherapy.

Table 3 summarizes studies in the United States that incorporate DNA flow cytometry measurements in node negative disease. One of the first studies, activated in 1983, was a prospective study evaluating cell kinetic parameters in fresh tissue as markers of prognosis in a group of node negative patients receiving no adjuvant

Table 3

Flow Cytometry And Breast Cancer In Clinical Trials

STUDY GROUP	PATIENTS #	COMMENTS (STUDY CHAIR)
SWOG 8366	N - (500)	Prospective. Nat'l Hx Study of Cell Kinetics and Prognosis. (McDivitt)
ECOG 7186 (INT0076)	N - (565)	Retrospective. Evaluation of ploidy and S phase to predict prognosis and response to chemotherapy (Dressler)
SWOG 9037 (INT0102)	N - (800)	Prospective. High S phase defines eligibility ER+ &PgR+, <2cm tumor. (Nat'l Hx for tumors too small for ER. (McGuire/Clark)

therapy. The results of this study are currently being analyzed by the Southwest Oncology Group (SWOG 8366). The first Intergroup study (INT0076) to look at these measurements was an Eastern Cooperative Oncology Group chaired study that was an ancillary laboratory study to the randomized clinical trial (INT011). This was a retrospective study, using formalin fixed paraffin embedded material from the primary tumor to measure ploidy and S phase to predict prognosis in the group of patients receiving no adjuvant therapy (11). It also allowed for the evaluation of these measurements to predict response to chemotherapy in the group of pa-tients randomized to either chemotherapy or observation. Preliminary results of this study were presented at the NIH consensus development conference on the Treatment of Early Stage Breast Cancer and will be discussed in more detail below (11).

The Intergroup trial (INT0102), chaired by the Southwest Oncology Group, uses SPF to determine eligi-bility for randomization in an otherwise good prognostic group (22). Patients whose ER+, PR+, less than 2cm in diameter tumors have a high S phase would be eligible for randomization to one of two treatment arms. Patients whose tumors have a low S phase (ER+, PR+, <2cm in diameter) would be assigned to observation only. Different cut points are used for diploid and aneuploid tumors to define high and low S phase. At this writing, no analyses have been published on this prospective dataset.

Let us discuss some of the data that has been published from the retrospective Intergroup trial (INT0076) that was presented at the NIH Consensus Conference (11). Figure 1 shows the relationship between ploidy status (DNA diploid vs DNA aneuploid) to predict for time to recurrence in the group of node negative pa

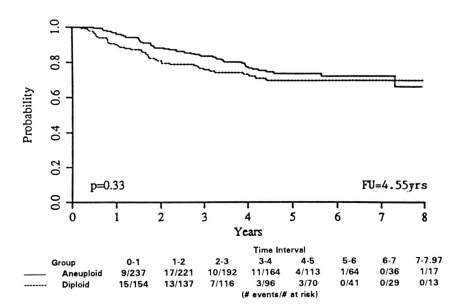

Group	0-1	1-2	2-3	Time Interval 3-4	4-5	5-6	6-7	7-7.97
——— Aneuploid	9/237	17/221	10/192	11/164	4/113	1/64	0/36	1/17
········ Diploid	15/154	13/137	7/116	3/96	3/70	0/41	0/29	0/13
				(# events/# at risk)				

Figure 1. Kaplan-Meier estimates of time to recurrence (TTR) by ploidy status in untreated patients who received no adjuvant therapy (surgery alone). Time interval represents number of events (recurrences) per number of patients at risk. No difference was observed between patients with diploid and aneuploid tumors in discriminating for time to recurrence. Reprinted with permission, JNCI Monographs, 1992.

88

tients who received no adjuvant therapy. Median time to
follow-up for this analysis was 4.5 years. We observe
that in this group of patients, ploidy status does not
discriminate for time to recurrence. Figure 2 shows the
relationship between proliferative capacity (SPF) to
predict time to recurrence. Using the median S-phase
value of 6.97% to define high and low S-phase groups, we
observe that patients whose tumors have a low S phase
have a significantly longer time to recurrence (11).
This was observed within the diploid as well as the
aneuploid group of tumors.

Table 4 summarizes those clinical trials in early
flow cytometry measurements. There are at least two

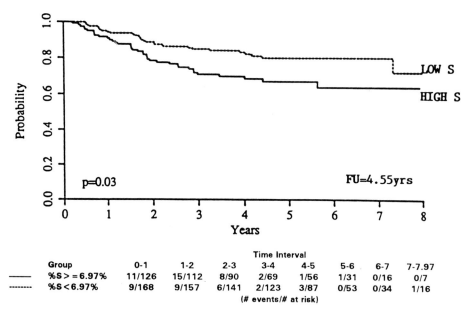

Time Interval								
Group	0-1	1-2	2-3	3-4	4-5	5-6	6-7	7-7.97
%S > = 6.97%	11/126	15/112	8/90	2/69	1/56	1/31	0/16	0/7
%S < 6.97%	9/168	9/157	6/141	2/123	3/87	0/53	0/34	1/16
				(# events/# at risk)				

Figure 2. Kaplan-Meier estimates of time to recurrence
(TTR) by percentage S phase in untreated patients. The
median S-phase value of the entire population of 6.97%
was used to define high (≥6.97%) and low (<6.97%) SPF.
Time interval represents number of events (recurrence)
per number of patients at risk. Patients whose tumors
had a low S phase had a significantly longer time to
recurrence compared to patients whose tumors had a high
SPF. P value is stratified for tumor size. Reprinted
with permission, JNCI Monographs, 1992.

stage and locally advanced disease that are using DNA
retrospective and two prospective studies evaluatingthe-
se measurements to predict response to therapy as well
as patient prognosis. The NSABP measures S phase and
ploidy in T1, T2, and T3 lesions from women undergoing
pre-surgical and post-surgical chemotherapy
(NSABP18.2) (23). In the group of patients undergoing
pre-surgical chemotherapy, both the pre-chemotherapy
biopsy sample and post-chemotherapy mastectomy sample
will be assayed for ploidy and S phase. This will allow
evaluation of these markers to predict response to
therapy in a unique prospective setting. The ECOG study
(EST4189) includes patients treated on five different
therapeutic regimens and compares these measurements in
primary lesions and corresponding lymph node metastases

Table 4

Flow Cytometry And Breast Cancer In Clinical Trials

STUDY GROUP	PATIENTS #	COMMENTS (STUDY CHAIR)
NSABP 18.2	T_1, T_2, T_3 ($>1,000$)	Prospective. Ploidy and S phase in pre-op and post-op chemotherapy. Prognosis and response to treatment. (Fisher)
ECOG 4189	N + (750)	Retrospective. Evaluation of ploidy and S phase to predict response to treatment and prognosis. (Dressler)
SWOG 8854 (INT0104)	N + (500) Post M	Prospective. ER+, PgR+, Evaluation of ploidy to predict response to treatment; prognosis, FCM/IMAGE. (Hermann)
NSABP B-04	N + (155) N - (243)	Retrospective. Evaluate ploidy and S phase to predict prognosis. Nat'l Hx Study. (Fisher)

(24). The SWOG chaired Intergroup Study (INT0104) is a prospective study evaluating ploidy status by both flow cytometry and image analysis as a marker to predict both response to treatment and patient prognosis in a group of postmenopausal node positive patients (25).

The NSABP study (B-04) was recently published and includes patients with clinically node positive and node negative disease with 10 year follow-up (21). This was a natural history study in which none of the patients received adjuvant chemotherapy. In this study, there was no correlation observed between ploidy status and the ability to predict either disease-free survival or OS. Measurement of SPF, however, showed a significantly longer disease-free survival and longer OS in patients whose tumors had a low S phase (21). This relationship was especially strong within the clinically node negative population.

Multivariate analysis performed in Study B-04 included both node positive and node negative patients. Variables tested included age, nuclear grade, ploidy, clinical node status, S phase and tumor size. When analyses were performed for the ability of these variables to predict disease-free survival, it was reported that only S phase and tumor size maintained their significance (S phase: $p=.04$; tumor size: $p<.001$). All of the other variables tested were not significant predictors in the multivariate setting (21). When the analyses was performed to predict for OS (21), tumor size was the single most significant factor ($p<.001$), with S phase approaching borderline significance ($p=.08$).

RESPONSE TO THERAPY

At this time, it is difficult to make any generalizations regarding the ability of these measurements to predict response to treatment. Investigators are just

beginning to evaluate ploidy and S phase in this setting. Several investigators who are evaluating ploidy
and S phase changes pre- and post-chemotherapy in locally advanced tumors from patients undergoing pre-surgical
chemotherapy have reported exciting results (14-19).
Brifford and colleagues reported that aneuploid tumors
showed a high rate of objective regression (14). A follow-up to this study further described that changes in
DNA histogram pattern observed throughout the course of
chemotherapy were associated with objective response,
but not risk of tumor recurrence (15). Remvikos and
colleagues reported that tumor responsiveness was related to a high S phase measured in the pretreatment specimen (16).

It is obvious that results from the clinical trials
and several other ongoing studies will better define how
this information can be used to predict response in the
individual breast cancer patient.

ACKNOWLEDGEMENTS

The author would like to acknowledge the many
individuals and institutions participating in the cooperative group trials and the helpful discussions and
information provided by Drs. Chester Hermann and Gary
Clark. This talk was dedicated to the memory of Dr.
William McGuire whose work in the field of breast cancer
continues to impact on the lives of patients, clinicians
and researchers worldwide.

REFERENCES

1.   Dressler LG, Bartow SA: DNA flow cytometry in
      solid tumors: Practical aspects and clinical appli-
      cations. Seminars in Diagnostic Pathology 6:55-82,
      1989.
2.   Coulson PB, Thornthwaite JR, Wooley TW, et al:
      Prognostic indicators including DNA histogram type,
      receptor content, and staging related to human
      breast cancer survival. Cancer Res. 44:4187-4196,

1984.

3. Kallioniemi OP, Hietanen T, Mattila J, et al:
    Aneuploid DNA content and high S phase fraction of
    tumor cells are related to poor prognosis in pa-
    tients with primary breast cancer. Eur. J. Clin.
    Oncol. 23:277-282, 1987.

4. Kallioniemi OP, Blanco G, Alavaikko M, et al:
    Improving the prognostic value of DNA flow cytome-
    try in breast cancer by combining DNA index and S
    phase fraction: A proposed classification. Cancer
    62:2183-2190, 1988.

5. Siguurdsson H, Baldetorp B, Borg A, et al: Indica-
    tors of prognosis in node negative breast cancer.
    New Eng. J. Med. 322:1045-1053, 1990.

6. Muss HB, Kute TE, Case LD, et al: The relationship
    of flow cytometry to clinical and biologic charac-
    teristics in women with node negative primary
    breast cancer. Cancer 64:1894-1990, 1989.

7. Clark GM, Owens MA, McGuire WL: A new S phase
    model predicts for recurrence for aneuploid as well
    as diploid node-negative breast cancer. Proc. ASCO
    10:44, 1991.

8. Winchester DJ, Duda RB, August CZ, et al: The
    importance of DNA flow cytometry in node negative
    breast cancer. Arch. Surg. 125:886-889, 1990.

9. Klintenberg C, Stal O, Nordenskjold, et al: Pro-
    liferative index, cytosolic estrogen receptors and
    axillary node status as prognostic predictors in
    human mammary carcinoma. Breast Cancer Res. and
    Treat. 7 (Suppl) 99-106, 1986.

10. O'Reilly SM, Camplejohn RS, Barnes DM, et al: Node
    negative breast cancer: Prognostic subgroups de-
    fined by tumor size and DNA flow cytometry. J.
    Clin. Oncol. 8:2040-2046, 1990.

11. Dressler LG, Eudey L, Gray R, et al: Prognostic
    potential of DNA flow cytometry measurements in
    node-negative breast cancer patients: Preliminary
    analysis of an Intergroup Study (INT0076). J.
    Natl. Cancer Inst. 11:167-172, 1992.

12. Ewers SB, Langstrom E, Baldetorp B, et al: Flow
    cytometric DNA analysis in primary breast carcino-
    mas and clinicopathological correlations. Cytome-
    try 5:408-419, 1984.

13. Beerman H, Klein Ph.M, Hermans J, et al: Prognos-
    tic significance of DNA-ploidy in a series of 690
    primary breast cancer patients. Int. J. Cancer
    45:34-39, 1990.

14. Brifford M, Spyratos F, Tubiana-Hulin M, et al:
    Sequential cytopunctures during pre-operative che-
    motherapy for breast cancer. Cytomorphologic chan-
    ges, initial tumor ploidy and tumor regression.
    Cancer 63:631-637, 1989.

15. Spyrotas F, Brifford M, Tubiana-Hulin M, et al:
    Sequential cytopunctures during pre-operative che-

motherapy for primary breast carcinoma. II. DNA
flow cytometry changes during chemotherapy, tumor
regression and short-term follow-up. Cancer 69:470-
475, 1992.

16. Remvikos Y, Beuzeboc P, Zadjela A, et al: Correlation of pretreatment proliferative activity with
response to cytotoxic chemotherapy. J. Natl. Cancer Inst. 81:1383-1387, 1989.

17. Seymour L, Bezwoda WR, Meyer K: Response to second
line hormone treatment for advanced breast cancer.
Predictive value of ploidy determination. Cancer
65:2720-2724, 1990.

18. Osborne CK: DNA flow cytometry in early breast
cancer. A step in the right direction. JNCI
81:1344-1345, 1989.

19. Dressler L, Mangalik A, Bartow SA, et al: The use
of DNA flow cytometry to characterize breast cancer
cells affected by pre-surgical (neoadjuvant) chemotherapy. Proc. AACR 33:213, 1992.

20. Lewis WE: Prognostic significance of flow cytometric DNA analyses in node negative breast patients.
Cancer 65:2315-2320, 1990.

21. Fisher B, Gunduz N, Costantino J, et al: DNA flow
cytometric analysis of primary operable breast
cancer. Relation of ploidy and S phase fraction to
prognosis of patients in NSABP B-04. Cancer
68:1465-1475, 1991.

22. Intergroup Trial (0102). SWOG 8897. ECOG 2188.
CALGB 8897. Phase III comparison of adjuvant chemotherapy with or without endocrine therapy in high
risk, node negative breast cancer patients, and a
natural history follow-up study in low risk, node
negative patients. Activated August 1989.

23. NSABP Protocol B-18. A "unified" trial to compare
short, intensive preoperative systemic adriamycin,
cyclophosphamide therapy with similar therapy administered in conventional postoperative fashion.
B-18.2. A study to evaluate DNA histograms by flow
cytometry.

24. Eastern Cooperative Oncology Group Laboratory Study
EST4189. Clinical significance and therapeutic
impact of DNA flow cytometry measurements of ploidy
and S phase and immunohistochemistry of the HER-
2NEU oncogene product and steroid hormone receptors: Prediction of disease progression and response to therapy in Stage II breast cancer patients.

25. Southwest Oncology Group Laboratory Study 8854.
Intergroup (INT0104). Prognostic value of cytometry measurements of breast cancer DNA from post
menopausal patients with involved nodes and receptor positive tumors: A comparison protocol to SWOG
8814.

8

PROCESSING OF SOLID TUMORS FOR DNA ANALYSIS BY FLOW CYTOMETRY

Charles L. Hitchcock, John F. Ensley, Mark Zalupski

INTRODUCTION

A review of the literature on flow cytometry for DNA studies of human tumor raises concerns regarding the wide range of results and the contradictory conclusions reached by these studies. Techniques used to produce a monodisperse suspension of cells or nuclei from a tumor are a principal source of this variability. Dissociation of either fresh or fixed, paraffin-embedded tissue requires disruption of cell-cell and cell-matrix attachments. Dissociation is easily accomplished for lymphatic and hematopoietic tissue but not for solid tumor tissue. Optimally, the dissociation technique should: (1) insure that the sample is representative of the tumor, (2) not induce excess background aggregates and debris (BAD), (3) optimize the cell yield per gram of tissue, and (4) maintain the desired phenotypes being studied. The ability to meet these goals varies with the tumor as well as with the dissociation technique. Individualization and validation of a dissociation technique, for either a fresh or fixed, paraffin-embedded tumor, is needed to meet these criteria. Understanding the impact of technical variables on a given tumor is essential.

SAMPLING

Sampling is the initial step in processing any

solid tumor for flow cytometry. The sample must provide
sufficient and representative material for analysis.
Heterogenous populations of cells within the tumor make
representative sampling of resected lesions or excisio-
nal biopsies difficult. This often necessitates sampling multiple sites from a solid tumor. Fresh tissue
should be quickly sampled and kept moist at 4°C. Areas
of grossly non-viable tumor, necrosis and fat must be
removed to minimize both debris and the possibility of
false DNA aneuploid peaks.

Fine needle aspiration (FNA) is a rapid and versatile in vivo and ex vivo technique for sampling specific
sites. It is a gentle mechanical dissociation method
that yields single cells and numerous cohesive plugs
rather than segments of tissue. FNA facilitates retention of cellular architecture that permits microscopic
correlation with flow cytometric or image analysis of
aspirates. FNA sampling can lead to increased diagnostic sensitivity and specificity (1,2).

The amount of diagnostic material in FNA samples
varies with the size, location, nature of the lesion,
number of passes taken, and experience and skill of the
person performing the aspircition. On-site cytological
assessment of the aspirate is essential for obtaining an
adequate amount of diagnostic material. Vindelov et al
(3) recommended using a hemocytometer to quantitate cell
yield per sample, and continue aspirating until 10^6
cells in 200 μl were obtained. The needle's contents
can be ejected into transportation or freezing media
(4,5), or directly into staining solution for flow
cytometry.

FNA and flow cytometry have been used together in
varying clinical settings. The combination has facilitated immunophenotyping in lymphadenopathies (6-8).
Testicular FNA samples have been used to define males at
increased risk for developing testicular carcinoma (9),

as well as for analysis of spermatogenesis (10,11). FNA has been found to be useful for sampling in vivo and ex vivo breast lesions for diagnosis (12-16) and for monitoring treatment response (17-19). Fuhr concluded that flow cytometric analysis of FNA samples is an excellent screening technique for tumors of the lung, breast, and liver (20,21), but that it is not useful for screening thyroid and pancreatic lesions.

DISSOCIATION OF FRESH SOLID TUMOR TISSUE

Dissociation of fresh solid tumors can be accomplished by chemical, mechanical, or enzymatic means, alone or in combination. In general, one gram of tissue contains 10^9 cells, and yet the best yields from dissociative techniques do not exceed 30% of these cells (22). Why? Cell yields vary with the tumor type and susceptibility of the tumor to physical and chemical treatments. Proliferating cells are considered to be the most susceptible to damage during dissociation (23). A dissociation technique should be individualized for the tumor being studied and evaluated for optimal and representative cell yields relative to the endpoint(s) being measured.

Enucleation Technique

Penetration of the DNA binding fluorochrome is facilitated by using enucleation techniques to render the cell and its nucleus permeable. These rapid techniques are especially applicable to small samples (3). The technique requires initial mechanical disruption of the tumor followed by lysis using a hypotonic or hypertonic solution, non-ionic detergent, or proteolytic digestion, alone or in combination. The resulting nuclear suspension is suitable for flow cytometric analysis of various nuclear antigens (24,25), or BrdU staining (26). It is not suitable for analysis of surface and cytoplasmic antigens, making it difficult to evaluate the suspension for loss of specific tumor cell populations from the sample.

Few studies have carefully evaluated enucleation techniques. Using a murine squamous cell carcinoma of the head and neck (SCCHN) model, we compared results using enucleation methods with those obtained with enzymatic and mechanical dissociation of the same tumor (27-29). In contrast to enzymatic and mechanical dissociation, we found that enucleation resulted in a 90% loss of nuclei, a preferential loss of cell subpopulations, and the presence of artifactual pseudo-hyperdiploid DNA populations. Risberg et al (30) demonstrated that not all enucleation methods give the same results. Using paired samples, they compared detergent lysis (31) with detergent/enzyme enucleation (32) and noted marked differences in DNA histogram parameters. The detergent/enzyme enucleation technique increased the release and detection of DNA aneuploid nuclei, while decreasing the following: G0/G1 peak coefficient of variation (CV), DNA index (DI), S phase fraction (SPF), and percent G2/M. Decreased SPF and G2/M percentages were attributed to decreased debris and aggregate formation following detergent/enzyme enucleation.

Techniques For Obtaining Whole Cell Suspensions

Formation of a suspension of single cells requires an alteration or destruction of the elements involved in tissue and cellular cohesion. This is accomplished by using one or more chemical, mechanical, or enzymatic techniques to obtain intact whole cells from a solid tumor. Because of tissue variability in extracellular matrix and cell type, there is no universal dissociation technique.

Chemical Dissociation

Chemical or chelation dissociation of fresh tissue requires disrupting or sequestering the Ca^{+2} and Mg^{+2} ions needed for maintenance of the cell-cell and cell-extracellular matrix adhesion (33). Ethylene-diaminoac-etate (EDTA) (34,35) or citrate ion (36) are chelating agents commonly used for this process. Tumor dissociation using chelating agents alone produces low cell yields, good retention of morphology, and poor clonogenicity (37). Thus, these agents are best used in conjunction with mechanical or enzymatic dissociation procedures.

Mechanical Dissociation

Mechanical dissociation breaks tissue up quickly to release cells that are loosely bound in the stroma. Methods used include repeated mincing either with scissors or sharp blades, homogenization, scraping the cut surface of the tumor, filtration through a nylon or steel mesh, vortexing, triturating with pipettes or with successively smaller gauge needles, or any combination of these. Used to dissociate solid tumors, these processes produce varying cell yield, cell viability, and preferential selection of DNA aneuploid cells (Table 1).

Enzymatic Dissociation

Enzymatic dissociation of solid tumors is often the technique of choice in studies requiring high yields with near normal morphology and function (37). A variety of enzymes have been used either alone or in combination to digest stromal elements as well as to break intercellular adhesions. These include: Non-specific proteases such as trypsin, pronase, papain, and pepsin; proteases specific for extracellular matrix components such as collagenase and elastase; and enzymes such as hyaluronidase that hydrolyze mucopolysaccharides. DNase

Table 1
Dissociation of Fresh Tumors

	N	Mechanical Dissociation (mean $\pm$ SD)	Enzymatic Dissociation (mean $\pm$ SD)	p-value
Colon Carcinoma				
Cell Yield	25	20.6 $\pm$ 4.1	37.5 $\pm$ 6.2	0.01
% Viability	25	49.7 $\pm$ 4.8	487.7 $\pm$ 2.5	0.01
G0/G1 Peak CV	25	3.8 $\pm$ 0.2	4.8 $\pm$ 0.4	0.01
Head and Neck Squamous Cell Carcinoma				
Cell Yield	77	14.1 $\pm$ 2.6	55.5 $\pm$ 6.1	0.0001
% Viability	77	46.2 $\pm$ 3.3	91.0 $\pm$ 1.3	0.0001
G0/G1 Peak CV	76	4.2 $\pm$ 0.16	4.8 $\pm$ 0.18	0.692

Total cell yield is 10^6 gram of tissue.
Cell viability was determined by dye exclusion.

is often used to hydrolyze the gel-like DNA-protein
complexes which entrap cells. The use of crude enzymes,
whose composition is often undefined (38) and varies
from lot to lot, has led to conflicting reports in the
literature on which enzyme is best used to dissociate a
particular tumor. For example, trypsin and collagenase
in their pure form, are ineffective in hydrolyzing
tissue. Their dissociative properties arise from variable concentrations of contaminating proteases, nucle-
ases, lipases, and polysaccharidases.

Evaluating Dissociation Methods

Evaluating dissociating techniques must take into
account all of the criteria specified above. Cell
yields vary with both the tumor and the dissociation
technique (28,29,39-43). In general, enzymatic dissociation results in increased cell yields, the significance of which depends on the tumor. When compared to
mechanical dissociation, enzymatic dissociation of

squamous cell (28,29) and colon carcinomas (44) results
in a significant increase in cell yields (Table 1).
This difference is not significant in cases of carcinomas of the breast, ovary, and lung (23,41), melanomas
(23,41) or sarcomas (40,42). Varying technical factors
such as the type or concentration of enzyme and duration
of digestion also alter cell yields (43,44). The ability to detect DNA aneuploid cells also varies with the
tumor type and dissociation protocol (Fig. 1-3). In
comparison to mechanical methods, enzymatic dissociation
can decrease the relative proportion of DNA aneuploid
cells obtained from different tumors (23,28,29,42-46).
This decrease can be attributed to an increased release
of inflammatory and stromal cells that function to
dilute the DNA aneuploid cell population to a point of
undetectability (Fig. 1,2) (23). The effect of tumor
type on the yield of subpopulations of tumor cells is
noted in our results (28,29,44,46). In contrast to
colon carcinoma (Fig. 2) and sarcomas (Fig. 3), we found
that mechanical dissociation of SCCHN tissue frequently
resulted in an apparent loss of DNA aneuploid cell
subpopulations compared to cell suspensions obtained by
enzymatic dissociation of the same tumor. In over 80%
of the tumors studied either a reduction or loss of DNA
aneuploid cells occurred (Fig. 1). Changing the conditions of enzymatic dissociation can also have a significant impact on the recovery of tumor cell subpopulations, as exemplified by the results with colon carcinoma
seen in Fig. 4, and by the studies of Bijman et al (47)
with SCCHN.

Evaluating Methods for Fixation of Dissociated Cells
     Fixation of dissociated cells and nuclei can yield
variable flow cytometric results. Commonly used fixatives include formalin, paraformaldehyde, glutaralde-
hyde, ethanol, methanol, and 20% acetic acid. Using

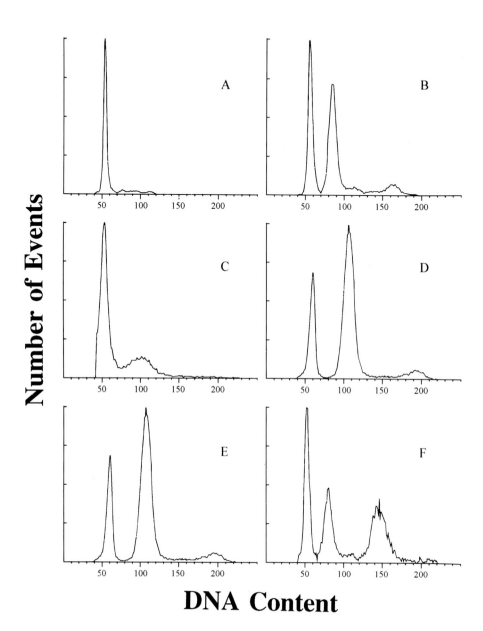

Figure 1. Paired DNA histograms of 3 different squamous cell carcinomas of the head and neck dissociated by mechanical (A, C and E) or enzymatic (B, D and F) means.

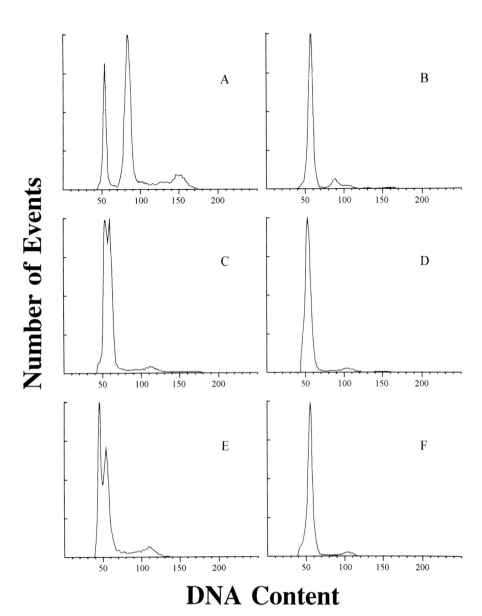

Figure 2. Paired DNA histograms from 3 different human colon carcinomas dissociated by mechanical (A, C and E) or enzymatic (B, D and F) means.

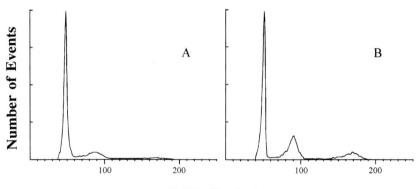

Figure 3. DNA histograms comparing mechanical (A) and enzymatic (B) dissociation of a human soft tissue sarcoma.

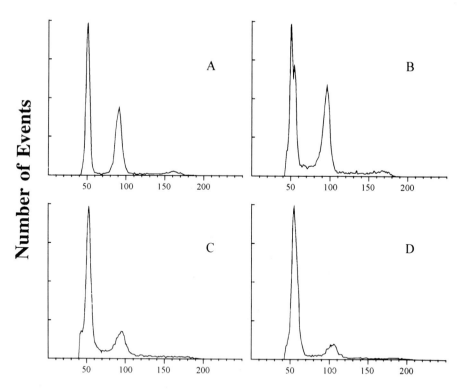

Figure 4. DNA histograms of the same human colon carcinoma comparing mechanical dissociation (A) with sequential samples digested with 0.1 mg/ml collagenase II for 30 min (B), 60 min (C), and 120 min (D).

different fixatives varies staining intensity (48,49)
and, thus, affects the quality of results. Also, fixation can decrease cell recovery (50). Our experience
with ethanol fixation indicates that using 50% pure
(Gold Shield) ethanol yields up to 100% cell recovery
after 60 min, compared to less than a 40% yield when
samples were fixed with 50% denatured ethanol (Table 2).
We also noted that fixation with denatured ethanol can
produce broad CVs and pseudo-DNA aneuploid peaks (Fig.
5). Because of these variations in DNA fluorochrome
binding stoichiometry, combining data from tissue treated by fixation techniques can distort results of a study
(27,28). Fixation can also lead to clumping, lysis, and
loss of cells. These effects can be minimized by vortexing the cell suspension while the fixative is being
added in a drop-wise manner.

Table 2
Effect of Ethanol Fixation on Cell Loss

| Duration of Fixation | Percentage of Cells Recovered Following EtOH Fixation | |
	Denatured (50%)	Gold Shield (50%)
10 Min	24	65
20 Min	36	80
30 Min	39	80
60 Min	36	100
120 Min	39	100
17 Hours	72	90
4 Days	69	Not Done
9 Days	100	Not Done

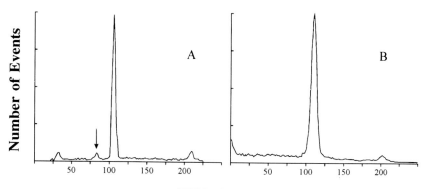

DNA Content

Figure 5. DNA histograms from paired samples of a colon carcinoma prepared fresh (A) and fixed and parrafin-embedded (B). The arrow indicates a small DNA diploid population in the fresh sample (A) that is lacking in the DNA histogram (B) from a section of fixed and parra-fin-embedded tumor (Koss and Wersto, personal communication).

DISSOCIATION OF SECTIONS OF FIXED, PARAFFIN-EMBEDDED TISSUE

Since its introduction in 1983 (96), use of "Hedl-ey's method", or a modification of it, to dissociate tumor nuclei from sections of fixed paraffin-embedded tissue has produced a plethora of journal articles. The technique facilitates retrospective DNA ploidy and/or cell proliferation studies of well-defined patient groups with known clinical follow-up (96-98) and makes accessible the clinical significance of these variables in rare or uncommon lesions (99-103). Although the use of fresh tissue is preferable, clinical use of this method of dissociating solid tumors is increasing.

PROBLEMS

The lack of standardized techniques for processing fixed, paraffin-embedded, or archival, tissue makes comparison of results difficult. This has been well demonstrated by the results of recent intra- and inter-laboratory studies (1,3,4). In general, results from

flow cytometric analyses of fixed, paraffin-embedded
tumors that tend to correlate are characterized by wide
peak CVs and increased background aggregates and debris
(BAD). The use of this archival material is also ham-
pered by inadequate internal standards for DNA ploidy
determination.

Peak Coefficient of Variation

When compared to fresh samples from the same tumor,
the G0/G1 peak of fixed samples may be asymmetric with
an increased (ie widened) coefficient of variation
(Table 3). These two effects taken together, can reduce
detection of a tumor cell subpopulation with near-diploid DNA content as a result primarily of fixation and
enzymatic digestion.

Table 3
Fresh vs Archival Tissue DNA Distribution
Variability CV, DI and %S

Tumor Type	Mean CV		Concordance[a]		Ref.
	Fresh	Archival	DI (r)	%S (r)	
Bladder	3.7	7.3	0.957	-	63
Breast	3.5	5.1	0.985	0.886	64
Lymphoma[b]	3.8	4.0	>0.95	0.940	65
Lymphoma[c]	4.4	4.5	-	-	65
Colon	3.6	5.0	0.988	0.588	66

[a] r represents coefficient of correlation
[b] Lymphomas stained with DAPI
[c] Lymphomas stained with PI

Background Aggregates and Debris (BAD)

BAD arise from necrotic tissue as well as from
fragments of cut nuclei and stroma. For paraffin block

sections, the amount of debris is further influenced by fixation, tissue cellularity, section thickness, and enzymatic digestion. Debris may obscure detection of DNA aneuploid cell populations present at low frequencies, and may contribute to artifactual DNA aneuploid or DNA tetraploid peaks. BAD form a continuum underlying the DNA histogram that affects the accuracy of cell-cycle analysis in an otherwise interpretable DNA histogram. Although software substraction algorithms can reduce their impact, in general, the SPF is higher from archival samples compared to fresh samples (67,68).

Lack of Controls

An external or internal DNA diploid reference cell is often needed to verify both the presence and location of a DNA aneuploid cell population in a DNA distribution. The effects of fixation and enzymatic digestion on DNA staining preclude the use of a reliable external standard and necessitate the use of an internal standard. The only reproducible internal standard is the non-malignant cell population found in a paraffin block containing both malignant and nonmalignant tissue. Both tissues can be separated and analyzed independently as well as admixed. (Please note: Blocks of normal tissue from another location or from another patient cannot be used as a standard) (69)

Fresh Versus Paraffin-Embedded Samples

Studies comparing DNA histograms derived from fresh and fixed, paraffin-embedded samples from the same tumor provide valuable information about the usefulness of this technique (Table 2). Differences in tumor DNA histograms can be attributed to sampling differences (70,71), tissue fixation (72), tissue dissociation (73), and DNA stain (74). In general, the percentage of DNA aneuploid nuclei in archival samples is lower than in

fresh tissue, but the detection of DNA aneuploid cell
populations is similar (Table 2). Concordance in SPF
values (Table 2) has been associated with the DNA ploidy
of the tumor cells and the method used for subtracting
BAD (68). Interlaboratory studies comparing DNA staining in paraffin sections from the same formalin-fixed
tissue suggest that technical variables contribute
significantly to differences in DNA histograms (59-62).

Initial Processing

The use of sections from fixed, paraffin-embedded
solid tumors for flow cytometric analysis requires
careful control of initial processing techniques.
Careful control begins with the tissue itself and its
sampling. Tissues vary in the effect that fixation has
on DNA staining and thus on debris and peak CVs (72,73).
As previously noted, a tumor must be sampled in such a
way as to remove necrotic and hemorrhagic tissue. The
use of multiple samples ensures detection of heterogenous cell populations in various tumors.

Fixation

Tissue fixation is a critical factor in determining
the quality of flow cytometric results. The fixative
used, as well as the size and density of the sample,
affect the intensity of DNA and nuclear protein staining
(48,77,78).

Different fixatives yield markedly different flow
cytometric results (Fig. 6) (52,79). Fixation with
Bouin's or Zenker's fixatives destroys antigen expression and often results in poor to uninterpretable DNA
histograms whereas more uniform results are commonly
obtained from tissue fixed in neutral-buffered formalin
(NBF). DNA histograms of varying quality are obtained
from tissue fixed in B5, ethanol, or Omnifix (Omni)
(78,79). DNA staining intensity is highest with NBF-

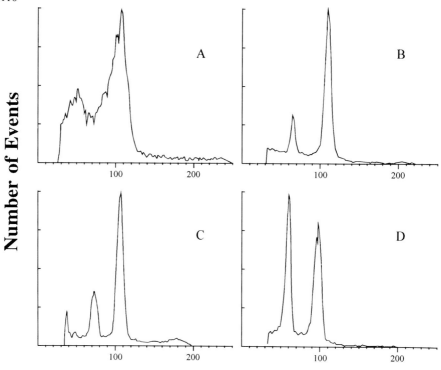

DNA Content

Figure 6. DNA histrograms from a colon carcinoma xenograft fixed in Bouin's fixative (A), neutral-buffered
formalin (B), OmniFix (C), and ethanol (D).

fixed tissue and lowest with Bouin's and Zenker's fixatives (80,81). Alanen et al (77) reported that the (DI)
was consistently larger, but not significantly so, in
as soon as possible after removal from the patient.
NBF-fixed samples compared to fresh or ethanol-fixed
samples from the same tumor. Fixation induces wide
G0/G1 peak CVs that may preclude detection of cell
populations with differences in DNA content less than
15% (DI ≤ 1.15) (82).

Fixation must be rapid to minimize tissue autolysis
that induces peaks with large CVs, false DNA aneuploid
peaks, increased debris, and artifactual antigen expression (77). Autolysis often continues during fixation,
because fixation tends to lag behind tissue penetration.
To minimize this effect, tissue samples should be less

than 2 cm square and less than 4 mm thick, with thinner
sections taken from fibrotic or sclerotic samples.
Samples should be placed in fixative at room temperature

Handling the Block

Sampling. Accurate assessment of a tumor's relative DNA content requires careful sampling of the paraffin block to insure adequate sample cellularity and
minimize unwanted necrotic and hemorrhagic material
(83,84). Sampling two or more blocks increases detection of tumor cell populations and agreement with results using fresh material (71,85,86). Normal cellular
components in the block must be present in the sample to
provide the necessary internal DNA diploid standard.

Sampling specific areas of interest begins with a
careful review of a stained section which helps to
identify specific areas to sample. A thin section
should also be taken at the end of sampling to determine
specimen adequacy. Specific areas can be obtained using
a skin-punch biopsy device (87), by scoring the block
face prior to sectioning, or by using a scalpel to cut
away unwanted tissue from a mounted rehydrated section
on a slide (88,89).

Sectioning. Sectioning of paraffin blocks may
produce debris by fragmenting nuclei. The amount of
debris is inversely related to both the thickness of the
section and the size of the nuclei (90,91). Using
sections 50 μm in thickness significantly reduces debris. The number of sections needed is dictated by the
cellularity of the specimen. In most cases, two sections yield ample material for processing.

Section Processing

Deparaffinization and rehydration. Deparaffinization and rehydration of the paraffin sections is

accomplished using standard histopathological techniques. Sections are deparaffinized in two changes of
xylene or a xylene substitute. Sections are then rehydrated to water or a saline solution through a series of
ethanol solutions of decreasing concentration. Most
protocols call for sections to be kept in each solution
for 10 to 60 min, and the process can be automated
(44,87,88). Thorough deparaffinization and rehydration
often leads to lower peak CVs.

Enzymatic Digestion

Pepsin digestion. Hedley's original protocol (51)
called for rehydrated sections to be digested with 0.5%
pepsin in acidified saline (pH 1.5) for 30 min. This
tissue dissociation step has been extensively modified,
often without examining the impact of the modifications
on results. A wide range of pepsin concentrations, from
0.05% to 1.0% have been used. DNA staining intensity
and peak CVs can differ substantially over this twenty-
fold variation (78,92,93) (Fig. 7). Retention of
antigenic epitopes on nuclear proteins are best retained
using low concentrations of pepsin (24,25). An increase
in digestion time tends to decrease peak CVs while
increasing staining intensity (Fig. 7), overall cell
yields, and the release of DNA aneuploid cells (44,92,
93). Other enzymes have been used, but results are not
as consistent as those obtained with pepsin (63-66,72,
81,84,94,95).

Methods of releasing nuclei. Tissue dissociation
and release of isolated nuclei is increased by shearing
the tissue. Several techniques can be used. Scissors
used to mince the tissue prior to or during digestion
can be useful for tenacious specimens. Most studies
report using vortexing at frequent intervals to increase
the yield of nuclei, but the yield is not as high as

A) PEAK CV

B) GO/GI: BEAD PEAK RATIO

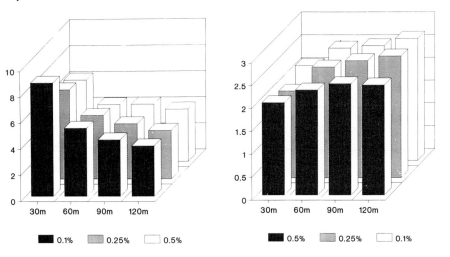

Figure 7. Effect of pepsin concentration and duration of digestion on (A) the coefficient of variation (CV) of DNA aneuploid G0/G1 peaks, and (B) the ratio of DNA aneuploid G0/G1 peaks to the peak of a known standard (DNA check beads).

that of more rigorous techniques. The highest yield of isolated nuclei is obtained from the increased shearing forces induced by repeated aspiration through a Pasteur pipette, a 23-27 gauge needle, or two Luer-locked syringes. High cell yields may be offset, however, by increased amounts of debris.

Stopping enzymatic digestion. The means to stop enzymatic digestion depends on the enzyme being used. Washes with media containing 10% fetal calf serum stops trypsin digestion. Pepstatin A or repeated washes with buffered media stop pepsin digestion.

Counting nuclei. The nuclei are washed in fresh media and counted with a hemocytometer or particle counter to insure an optimal concentration of $1-2 \times 10^6$ nuclei/ml for subsequent DNA staining and analysis.

DNA Staining

DNA staining influences the quality of DNA histograms. Many studies have modified Hedley's original staining method by using different DNA fluorochromes and staining protocols, but only a few studies have examined the impact of these modifications on results.

DNA fluorochromes. Propidium iodide (PI), ethidium bromide (EB), and 4,6-diamidino-2-phenylindole (DAPI) are the most commonly used DNA-binding fluorochromes for staining nuclei released from archival material. Staining is saturable with all three fluorochromes (51,96) but the intensity of staining is less for fixed nuclei than for fresh nuclei at saturating concentrations.

Staining conditions for PI. Propidium iodide is the most widely used DNA-binding fluorochrome. It has been dissolved in salt solutions of varying tonicity and make-up (44,78,97) and in complex solutions such as Vindelov's solution (32). Studies have not been done to determine whether there is an optimal solution. Detergents are included to further solubilize the cytoplasm, while spermine or polyethylene glycol are used to stabilize DNA.

Cell concentration for PI staining is an important consideration. Optimal staining with PI is obtained at a cell concentration of 2-3 x 10^6 cells/50 μg/ml of PI (96). Saturation of DNA by PI decreases as the cell concentration increases. PI staining is done in the dark for a minimum of 1 h at 4°C. There no apparent difference in results when nuclei are stained for 1 h up to 24 h (88).

RNase Treatment. RNase treatment is used to eliminate the artifactual broadening of the DNA content caused by the PI staining of ds-RNA. This treatment can be accomplished prior to, concurrently with, or after PI staining. Because there is no standardization of this step, several points must be considered. The highest quality RNase must be used to minimize the presence of contaminating proteolytic enzymes and DNase (98). It should be noted that RNase activity is inhibited by many metallic cations including Mg^{+2} at concentrations greater than 10 mM, and that the optimal pH for RNase is 6.0 to 7.5. Digestion can be carried out at $37^{\circ}C$ or at room temperature without significant alterations in results.

Storage. Samples stored after rehydration, enzymatic dissociation, or DNA staining make batching easier. Storing sections or nuclei at $4^{\circ}C$ in ethanol, distilled water, or saline may well improve the quality of DNA histograms and decrease intersample staining variation (53,99). Prolonged freezing is also possible without deleterious effects (100). Overnight storage, at $4^{\circ}C$, of the stained nuclei may improve the quality of the DNA histograms by allowing for complete equilibrium or access to DNA by the dye molecules. Stained nuclei can be kept at $4^{\circ}C$ for several days without staining degradation. This can lead to increased numbers of aggregates and clumps of nuclei that can be disrupted by repeated aspiration or by sonication (101).

SUMMARY

The future of flow cytometric analysis of solid tumor samples lies in reproducibility. At present, there is no single optimum or standardized technique for processing fresh or fixed, paraffin-embedded solid tumor samples for flow cytometry. The technical variables for processing solid tumors for subsequent flow cytometric analysis must be optimized and standardized within each laboratory. The goal of this manuscript has been to provide a better understanding of the problems inherent

116

in these techniques.

ACKNOWLEDGEMENTS

This work was supported in part by grants from the
American Registry of Pathology, Washington, D.C., RO3
CA53293 and UO1 CA62809; and, could not have been under-
taken without the technical assistance of Dr. Joe
Griffin, Annette Geisel, Karen Scott, Zosia Maciorowski
and Halina Pietraszkiewicz.

REFERENCES

1. Joensuu H, Klemi P, Eerola E: Diagnostic value of
    flow cytometric DNA determination combined with
    fine needle aspiration biopsy in thyroid tumors.
    Analyt. Quant. Cytol. Histol. 9:328-333, 1987.
2. Joensuu H, Klemi P, Eerola E: Flow cytometric DNA
    analysis combined with fine needle aspiration biop-
    sy in the diagnosis of palpable metastases. Anal.
    Quant. Cytol. Histol. 10:256-260, 1988.
3. Vindelov LL, Christensen IJ: A review of tech-
    niques and results obtained in one laboratory by an
    integrated system of methods designed for routine
    clinical flow cytometric DNA analysis. Cytometry
    11:753-770, 1990.
4. Stone KR, Craig RB, Palmer JO, et al: Short-term
    cryopreservation of human breast carcinoma cells
    for flow cytometry. Cytometry 6:357-361, 1985.
5. Vindelov LL, Christensen IJ, Keiding N, et al:
    Long-term storage of samples for flow cytometric
    DNA anaysis. Cytometry 3:317-322, 1982.
6. Hanson CA, Schnitzer B: Flow cytometric analysis
    of cytologic specimens in hematologic disease. J.
    Clin. Lab. Anal. 3:207, 1989.
7. Katz RL, Gritsman A, Cabanillas F, et al: Fine
    needle aspiration cytology of peripheral T-cell
    lymphoma. A cytologic, immunologic, and cytometric
    study. Am. J. Clin. Pathol. 91:120-131, 1990.
8. Johnson A, Akerman M, Cavallin-Stahl E: Flow cyto-
    metric detection of B-clonal excess in fine needle
    aspirates for enhanced diagnostic accuracy in non-
    Hodgkin's lymphoma. Histopathology 11:581-590,
    1987.
9. Nagler HM, Kaufman DG, O'Toole KM, Sawczuk IS:
    Carcinoma in situ of the testes: Diagnosis by aspi-
    ration flow cytometry. J. Urol. 143:359-361, 1990.
10. Kaufman DG, Nagler HM: Aspiration flow cytometry
    of the testes in the evaluation of spermatogenesis
    in the infertile male. Fertil. Steril. 48:287-291,

1987.

11. Skoog SJ, Evans CP, Hayward IJ, et al: Flow cytometry of fine needle aspirations of the S-D rat testis: Defining normal maturation and the effects of multiple biopsies. J. Urol. 146:620-623, 1991.

12. Greenebaum E, Koss LG, Sherman AB, Elequin F: Comparison of needle aspiration and solid biopsy techniques in the flow cytometric study of DNA distributions of surgically resected tumors. Am. J. Clin. Path. 82:559-564, 1984.

13. Shabot MM, Goldberg IM, Schick P, et al: Aspiration cytology is superior to tru-cut needle biopsy in establishing the diagnosis of clinically suspicious breast masses. Ann. Surg. 196:122-126, 1982.

14. Lykkesfeldt AE, Balslev I, Christensen IJ, et al: DNA ploidy and S-phase fraction in primary breast carcinomas in relation to prognostic factors and survival for premenopausal patients at high risk for recurrent disease. Acta Oncol. 27:749-756, 1988.

15. Palmer JO, McDivitt RW, Stone KR, et al: Flow cytometric analysis of breast needle aspirates. Cancer 62:2387-2391, 1988.

16. Fallenius AG, Askensten UG, Skoog LK, Auer GU: The reliability of microspectrophotometric and flow cytometric nuclear DNA measurements in adenocarcinoma of the breast. Cytometry 8:260-266, 1987.

17. Magdelenat H, Merle S, Zajdela A: Enzyme immunoassay of estrogen receptors in fine needle aspirates of breast tumors. Cancer Res. 46(suppl):4265s-4267s, 1986.

18. Briffod M, Spryratos F, Tubiana-Hulin M, et al: Sequential cytopunctures during preoperative chemotherapy for primary breast carcinoma, morphologic changes, initial tumor ploidy, and tumor regression. Cancer 63:631-637, 1989.

19. Remvikos Y, Beuzeboc P, Zajdela A, et al: Correlation of pretreatment proliferative activity of breast cancer with the response to cytotoxic chemotherapy. J. Natl. Cancer Inst. 81:1383-1387, 1989.

20. Fuhr JE: Flow cytometry in clinical oncology. Diagnostics Clin. Testing 28:24-29, 1990.

21. Fuhr JE, Nelson HS, Kattine AA: Analysis of flow cytometry of liver fine needle aspirates. Cytometry 6(suppl):35, 1993.

22. Pallavicini M: Solid tissue dispersal for cytokinetic analyses. In: Techniques in Cell Cycle Analysis, J Gray, Z Darzynkiewicz (eds), New York, Humana Press, pp 139-162, 1986.

23. Costa A, Silvestrini R, Del Bino G, Motta R: Implications of disaggregation procedures on biological representation of human solid tumors. Cell Tissue Kinet. 20:171-180, 1987.

24. Clevenger CV, Epstein AL, Bauer KD: Quantitative

analysis of a nuclear antigen in interphase and mitotic cells. Cytometry 8:280-286, 1987.

25. Lincoln ST, Bauer KD: Limitations in the measurement of c-myc oncoprotein and other nulear antigens by flow cytometry. Cytometry 10:456-462, 1989.

26. Ohyama S, Yonenura Y, Miyazaki I: Prognostic value of S-phase fraction and DNA ploidy studied with in vivo administration of bromodeoxyuridine on human gastric cancers. Cancer 65:116-121, 1990.

27. Ensley JF, Maciorowski Z, Hassan M, et al: The potential and pitfalls of solid tumor flow cytometry with respect to squamous cell cancers of the head and neck. In: Head and Neck Oncology Research. Proceedings of the Second International Research Conference on Head and Neck Cancer, GT Wolf, T Carey, (eds), Kugler Publishers, Amsterdam, Holland, pp 213-224, 1988.

28. Ensley JF, Maciorowski Z, Hassan M, et al: Characterization and clinical correlation of cellular DNA parameters in untreated and recurrent squamous cell cancers of the head and neck. Cytometry 10:334-338, 1989.

29. Ensley JF, Maciorowski Z, Pietraszkiewicz H, et al: Solid tumor preparation for flow cytometry using a standard murine model. Cytometry 8:488-493, 1987.

30. Risberg B, Stal O, Eriksson L-L, Hussein A: DNA flow cytometry on breast carcinomas: Comparison of a detergent and enzyme-detergent preparation method. Anal. Cell. Pathol. 2:287-295, 1990.

31. Thornthwaite JT, Sugarbaker EV, Temple WJ: Preparation of tissues for DNA flow cytometric analysis. Cytometry 1:229-237, 1980.

32. Vindelov LL, Christensen IJ, Nissen NI: A detergent-trypsin method for the preparation of nuclei for flow cytometric DNA analysis. Cytometry 3:323-327, 1983.

33. Berwick L, Corman DR: Some chemical factors in cellular adhesion and stickiness. Cancer Res. 22:982-986, 1962.

34. Oldbring J, Hellsten S, Lindholm K, et al: Flow DNA analysis in the characterization of carcinoma of the renal pelvis and ureter. Cancer 64:2141-2145, 1989.

35. Weiser M: Intestinal epithelial cell surface membrane glycoprotein synthesis. I. An indicator of cellular differentiation. J. Biol. Chem. 248:2536-2541, 1973.

36. Koss LG, Wolley RC, Schreiber K, Mendecki J: Flow-microflurometric analysis of nuclei isolated from various normal and malignant human epithelial tissues. A preliminary report. J. Histochem. Cytohem. 25:565-572, 1977.

37. Pallavicini M: Solid tissue dispersal for cytokinetic analyses. In: Techniques in Cell Cycle

Analysis, J Gray, Z Darzynkiewicz (eds), New York, Humana Press, pp 139-162, 1986.

38. Bashor MM: Dispersion and disruption of tissues. Methods Enzymol. 58:119-131, 1979.
39. McDivitt RW, Stone KR, Meyer JS: A method for dissociation of viable human breast cancer cells that produces flow cytometric kinetic information similar to that obtained by thymidine labeling. Cancer Res. 44:2628-2633, 1984.
40. Rous P, Jones FS: A method for obtaining suspensions of living cells from the fixed tissues and for the plating of individual cells. J. Exp. Med. 23:549-555, 1916.
41. Slocum HK, Pavelic ZP, Rustum YM, et al: Characterization of cells obtained by mechanical and enzymatic means from human melanoma, sarcoma and lung tumors. Cancer Res. 41:1428-1434, 1981.
42. Chassevent A, Daver A, Bertrand G, et al: Comparative flow DNA analysis of different cell suspensions in breast carcinoma. Cytometry 5:263-267, 1984.
43. Zalupski M, Ensley J, Ryan J, et al: Comparative dissociation techniques and flow cytometry in soft tissue and osteogenic neoplasms. Proc. AACR 30: 233, 1989.
44. Crissman JD, Zarbo RJ, Niebylski CD, et al: Flow cytometric DNA analysis of colon adenocarcinomas: A comparative study of preparatory techniques. Modern Pathology 1:198-204, 1988.
45. Ljung B-M, Mayhall B, Lottich C, et al: Cell dissociation techniques in human breast cancer - variations in tumor cell viability and DNA ploidy. Breast Cancer Res. Treat. 13:153-159, 1989.
46. Ensley JF, Maciorowski Z, Hassan M, et al: The loss of DNA aneuploid cells during tumor dissociation in human colon and head and neck cancers analyzed by flow cytometry. Cytometry 14:550-558, 1993.
47. Bijman JTh, Wagener DJTh, van Renned H, et al: Flow cytometric evaluation of cell dispersion from human head and neck tumors. Cytometry 6:334-341, 1985.
48. Becker RL, Mikel UV: Interrelation of formalin fixation, chromatin compactness and DNA values as measured by flow cytometry and image cytometry. Anal. Quant. Cytol. Histol. 12:333-341, 1990.
49. Holtfreter HB, Cohen N: Fixation-associated quantitative variations of DNA fluorescence observed in flow cytometric analysis of hemopoietic cells from adult diploid frogs. Cytometry 11:676-685, 1990.
50. Petersen SE: Flow cytometry of human colorectal tumors: Nuclear isolation by detergent technique. Cytometry 6:452-460, 1985.
51. Hedley DW, Friedlander ML, Taylor IW, et al: Meth-

od for analysis of cellular DNA content of paraffin-embedded pathological material using flow cytometry. J. Histochem. Cytochem. 31:1333-1335, 1983.

52. Hedley DW: Flow cytometry using paraffin-embedded tissue: Five years on. Cytometry 10:229-241, 1989.

53. Merkel DE, McGuire WL: Ploidy, proliferative activity and prognosis. DNA flow cytometry of solid tumors. Cancer 65:1194-1205, 1990.

54. Hitchcock CL, Norris HJ, Khalifa MA, Wargotz ES: Flow cytometric analysis of granulosa tumors. Cancer 64:2127-2132, 1989.

55. Hitchcock CL, Norris HJ: Flow cytometric analysis of endometrial stromal sarcomas. Am. J. Clin. Path. 97:267-271, 1992.

56. Seidman JD, Berman JJ, Hitchcock CL, et al: DNA analysis of cardiac myxomas: Flow cytometry and image analysis. Human Pathol. 22:495-500, 1991.

57. Wenig B, Hitchcock C, Ellis G, Gnepp D: Metastasizing mixed tumor of salivary glands (MMTSG): A clinicopathologic and flow cytometric analysis. Lab. Invest. 64:66A, 1991.

58. Wenig B, Hitchcock C, Griffin J, Heffner D: Olfactory esthesioneuroblastoma: Support for a grading system based on flow cytometric and histopathologic correlation. Lab. Invest. 64:66A, 1991.

59. Coon JS, Deitch AD, de Vere White RW, et al: Interinstitutional variability in DNA flow cytometric analysis of tumors. The National Cancer Institute's Flow Cytometry Network Experience. Cancer 61:126-130, 1988.

60. Wheeless LL, Coon JS, Cox C, et al: Measurement variability in DNA flow cytometry of replicate samples. Cytometry 10:731-738, 1989.

61. Hitchcock CL: (for "The Flow Cytometry Inter-Laboratory Study Group"). Inter-laboratory analysis of DNA by flow cytometry. Cytometry suppl 4:82, 1990.

62. Hitchcock CL: Variability in flow cytometric results using identical archival samples. Cytometry suppl 5:46, 1991.

63. Jacobsen AB, Fossa SD, Thorud E, et al: DNA flow cytometric values in bladder carcinoma biopsies obtained from fresh and paraffin-embedded material. APMIS 96:25-29, 1988.

64. Wingren S, Hatschek T, Stal O, et al: Comparison of static and flow cytometry for estimation of DNA index and S-phase fraction in fresh and paraffin-embedded breast cancer tissue. Acta Oncol. 6:793-797, 1988.

65. Camplejohn RS, Macartney JC, Morris RW: Measurement of S-phase fractions in lymphoid tissues comparing fresh versus paraffin-embedded tissue and 4',6'-diamidino-2 phenylindole dihydrochloride versus propidium iodide. Cytometry 10:410-416, 1989.

66. Schutte B, Reynders MMJ, Bosman FT, Blijham GH: Flow cytometric determination of DNA ploidy level in nuclei isolated from paraffin-embedded tissue. Cytometry 6:26-30, 1985.

67. Rabinovitch PS: Numerical compensation for the effects of cell clumping on DNA content histograms. Cytometry Suppl 4:27, 1990.

68. Weaver DL, Bagwell CB, Hitchcox SA, et al: Improved flow cytometric determination of proliferative activity (S-phase fraction) from paraffin-embedded tissue. Am. J. Clin. Pathol. 94:576-584, 1990.

69. Dressler LG, Bartow SA: DNA Flow cytometry in solid tumors: Practical aspects and clinical applications. Seminars Surg. Pathol. 6:55-82, 1989.

70. Frierson HF Jr: Flow cytometric analysis of ploidy in solid neoplasms: Comparison of fresh tissues with formalin-fixed paraffin-embedded specimens. Human Pathol. 19:290-294, 1988.

71. Owainati AAR, Robins RA, Hinton C, et al: Tumor aneuploidy, prognostic parameters and survival in primary breast cancer. Br. J. Cancer 55:449-454, 1987.

72. Pelstring RJ, Hurtubise PE, Swerdlow SH: Flow-cytometric DNA analysis of hematopoietic and lymphoid proliferations: A comparison of fresh, formalin-fixed and B5-fixed tissues. Human Pathol. 21:551-558, 1990.

73. Rabinovitch PS, Reid BJ, Haggitt RC, et al: Progression to cancer in Barrett's Esophagitis is associated with genomic instability. Lab. Invest. 60:65-71, 1988.

74. McIntire TL, Goldey SH, Benson NA, Braylan RC: Flow cytometric analysis of DNA in cells obtained from deparaffinized formalin-fixed lymphoid tissues. Cytometry 8:474-478, 1987.

75. Kallioniemi O-P: Comparison of fresh and paraffin-embedded tissue as starting material for DNA flow cytometry and evaluation of intratumor heterogeneity. Cytometry 9:164-169, 1988.

76. McIntire TL, Murphy WM, Coon JS, et al: The prognostic value of DNA ploidy combined with histologic substaging for incidental carcinoma of the prostate gland. Am. J. Clin. Pathol. 89:370-373, 1988.

77. Alanen KA, Joensuu H, Klemi PI: Autolysis is a potential source of false aneuploid peaks in flow cytometric DNA histograms. Cytometry 10:417-425, 1989.

78. Herbert DJ, Nishiyama RH, Bagwell CB, et al: Effects of several commonly used fixatives on DNA and total nuclear protein analysis by flow cytometry. Am. J. Clin. Pathol. 91:535-541, 1989.

79. Esteban JM, Sheibani K, Owens M, et al: Effects of various fixatives and fixation conditions on DNA

ploidy analysis. A need for strict internal DNA standards. Am. J. Clin. Pathol. 95:460-466, 1991.

80. Hostetter AL, Hrafnkelsson J, Wingre SOW, et al: A comparative study of DNA cytometry methods for benign and malignant thyroid tissue. Am. J. Clin. Pathol. 89:760-763, 1988.

81. Emdin SO, Stenling R, Roos G: Prognostic value of DNA content in colorectal carcinoma. A flow cytometric study with some methodologic aspects. Cancer 60:1282-1287, 1987.

82. van den Ingh HF, Griffien G, Cornelisse CJ: Letter to the editor: DNA aneuploidy in colorectal adenomas. Br. J. Cancer 55:351, 1987.

83. Kute TE, Gregory B, Galleshaw J, et al: How reproducible are flow cytometry data from paraffin-embedded blocks? Cytometry 9:494-498, 1988.

84. Mesker WE, Eysackers MJ, Ouwerkerk-van Velzen MCM, et al: Discrepancies in ploidy determination due to sampling errors. Anal. Cytol. Pathol. 1:87-95, 1989.

85. Beerman H, Smit VTHBM, Kluin PM, et al: Flow cytometric analysis of DNA stemline heterogeneity in primary and metastatic breast cancer. Cytometry 12:147-154, 1991.

86. Quirke P, Dixon M, Clayden AD, et al: Prognostic significance of DNA aneuploidy and cell proliferation in rectal adenocarcinomas. J. Pathol. 151: 285-291, 1987.

87. Lundberg S, Carstensen J, Rundquist I: DNA flow cytometry and histopathological grading of paraffin-embedded prostate biopsy specimens in a survival study. Cancer Res. 47:1973-1977, 1987.

88. Sickle-Santanello BJ, Farrar WB, DeCenzo JF, et al: Technical and statistical improvements for flow cytometric DNA analysis of paraffin-embedded tissue. Cytometry 9:594-599, 1988.

89. Murad T, Bauer K, Scarpelli DG: Histopathologic and flow cytometric analysis of adenomatous colonic polyps. Arch. Pathol. Lab. Med. 113:1003-1008, 1989.

90. Camplejohn RS: Comments on "Effects of section thickness on quality of flow cytometric DNA content determination in paraffin-embedded tissues". Cytometry 7:612-615, 1986.

91. Stephenson RA, Gay H, Fair WR, Melamed MR: Effect of section thickness on quality of flow cytometric DNA content determinations in paraffin-embedded tissues. Cytometry 7:41-44, 1986.

92. Hitchcock CL, Scott K: Optimization of techniques for flow cytometric analysis of DNA from archival material. Cytometry Suppl 4:102, 1990.

93. Hitchcock C, Scott K: Flow cytometric analysis of archival tissue: The use of a model tumor system to examine the effects of technical parameters. Lab.

Invest. 64:122A, 1991.

94. Albro J, Bauer KD, Hitchcock CL, Wittwer CT: Improved DNA content histograms from formalin-fixed, paraffin-embedded liver tissue by proteinase K digestion. Cytometry 14:673-678, 1993.

95. van Driel-Kulker AMJ, Mesker WE, van Velzen I, et al: Preparation of monolayer smears from paraffin-embedded tissue for image cytometry. Cytometry 6:268-272, 1985.

96. Schutte B, Reynders MMJ, Bosman FT, Blijham GH: Flow cytometric determination of DNA ploidy level in nuclei isolated from paraffin-embedded tissue. Cytometry 6:26-30, 1985.

97. Shapiro HM: Flow cytometry of DNA content and other indicators of proliferative activity. Arch. Pathol. Lab. Med. 113:591-597, 1989.

98. Deitch AD, Law H, DeVere White R: A stable propidium iodide staining procedure for flow cytometry. J. Histochem. Cytochem. 30:967-972, 1982.

99. McLemore DD, El Naggar A, Stephens LC, Jardine JH: Modified methodology to improve flow cytometric DNA histograms from paraffin-embedded material. Stain Technology 65:279-291, 1990.

100. Morkve O: Long-term storage of nuclear suspensions prepared from paraffin-embedded material for flow cytometric DNA analysis. Anal. Cell. Pathol. 2:327-331, 1990.

101. Gonchoroff NJ, Ryan JJ, Kimlinger TK, et al: Effect of sonication on paraffin-embedded tissue preparation from DNA flow cytometry. Cytometry 11:642-646, 1990.

9

SJL MODEL FOR LYMPHOMA TREATMENT

M. Cankovic, T. Wrone-Smith, E. VanBuren and S. Lerman

INTRODUCTION

Aged SJL mice spontaneously develop a B-cell malignancy, which has been compared with human follicular center cell lymphoma, an inevitably fatal disease despite conventional therapy (1-4). SJL lymphomas provide a model for development of a new treatment for this disease by revealing the existence of a novel target for therapy, nonmalignant lymphocytes upon which the tumor cells are dependent for growth (5).

Maximum growth of SJL lymphomas only occurs in the presence of syngeneic CD4+ T-lymphocytes, proliferation of which is stimulated by tumor cells in a mixed lymphocyte tumor cell interaction (MLTI) (6,7). In turn, it has been shown that in vitro proliferation of tumor cells is dependent upon lymphokines secreted by activated syngeneic CD4+ T-lymphocytes (8). Therefore, the CD4+ cell-dependence of SJL tumor cells can serve as a target for therapy (9,10). We have reported that SJL mice bearing a transplantable lymphoma treated with 100 mg/kg of cyclophosphamide (Cy) ip (RCS/Cy), survived significantly longer than untreated tumor-bearing mice (RCS5 mice) (11). The efficacy of this treatment was derived predominantly from decreased proliferation of CD4+ T-lymphocytes in response to tumor cells, which resulted, in part, from the suppressive action of a population of CD8+ T-lymphocytes (11).

The ensuing study was undertaken to test three predictions. First, given the fact that CD4+ T-lymphocytes from RCS/Cy mice proliferate poorly _in vitro_ in response to irradiated syngeneic tumor cells, we predicted that _in situ_ tumor-driven activation/proliferation of CD4+ cells would be inferior in RCS/Cy compared with RCS5 mice. Therefore, we sought to assess numbers and percentages of activated/proliferating CD4+ T-lymphocytes during the protracted and rapid courses of tumor growth, respectively, in RCS/Cy and in RCS5 mice. Secondly, based upon the prolonged survival of RCS/Cy mice, we sought to test the prediction that a smaller fraction of tumor cells would be in cell cycle in RCS/Cy compared with RCS5 mice.

Lastly, we sought to better define the CD8+ T-lymphocyte subpopulation that was responsible for specific suppression of CD4+ cell responsiveness in the MLTI. Appearance of a population of activated and/or proliferating CD8+ T-lymphocytes in RCS/Cy mice coincident with the development of the capacity of CD8+ T-lymphocytes to suppress the MLTI would be consistent with the suppressive subpopulation being contained within this activated and/or proliferating fraction.

Therefore, multi-color flow microfluorometry was used to monitor both cell cycle distribution and the expression of activation markers on CD4+ and CD8+ T-lymphocytes in RCS5 and RCS/Cy mice during the course of tumor growth. The activation markers chosen for this study were CD25 and CD44. CD25 defines the α chain of the IL-2 receptor, presence of which in conjunction with the constitutively expressed ß chain forms the high affinity IL-2 receptor expressed after T-cell activation (12). CD44 (formerly designated Pgp-1) is a cell surface adhesion/activation molecule, high expression of which distinguishes antigen-primed or memory T-cells from naive T-cells (13,14).

The ensuing results support the predictions which formed the premise for this study. Furthermore, the results better define this murine model for chemoimmunotherapy.

MATERIALS AND METHODS
Mice and Tumors

Five-week-old female SJL/J mice were purchased from the Jackson Laboratories (Bar Harbor, ME). The properties and history of the RCS5 tumor have been described previously (15). The tumor was transplanted weekly by iv injection of 10^7 highly tumorous lymph node cells. When noted, tumorous lymphoid cells were enriched (final purity averaged >90%) for tumor cells by antibody and C depletion of T-cells (15).

Cy Therapy and Experimental Design

Mice were injected iv with 10^7 highly tumorous lymph node cells and, where indicated, injected ip with 100 mg/kg of Cy (Cytoxan, Mead-Johnson, Bristol Myers Squibb Oncology Division, Syracuse, NY) one day after tumor cells (RCS/Cy treatment). On designated days, 3 RCS/Cy or RCS5 mice were euthanized, and spleens or selected lymph nodes (cervical, inguinal, brachial, axillary, mesenteric and sacral) were removed. On any given day on which samples were obtained, a normal SJL mouse was also euthanized, and its lymphoid cells were subjected to the same analytical procedures.

Immunofluorescence

Cell suspensions were prepared, erythrocytes were lysed by a rapid water treatment, and counts were made. Cells were stained by modifications of previously described procedures (16). Briefly, for IL-2 receptor expression, lymphoid cells were stained directly with 7D4-fluorescein isothiocyanate (FITC) (PharMingen, San

Diego, CA) and with anti-CD8 (YTS 169.4)-phycoerythrin (PE) (Coulter Corp., Hialeah, FL) or anti-CD4 (clone GK 1.5)-PE (Becton Dickinson, Mountainview, CA). For CD44 expression, cells were stained directly with anti-CD44 (clone IM7)-FITC (PharMingen) and anti-CD4 or anti-CD8 as described above. Numbers of RCS5 lymphoma cells were determined by reacting lymphoid suspensions with anti-H-2D^S (clone 20-8-4s) followed by staining with the secondary reagent, goat anti-mouse IgG-PE (Tago, Burlingame, CA). Cells were identified as tumor cells by being nonreactive with anti-H-2D^S and anti-IgG-PE. This approach has been used in the past and is dependent upon the fact that RCS5 lymphoma cells are surface Ig negative and lack expression of the class I MHC antigen, H-2D^S (16). In all cases, isotype or secondary reagent controls were used to distinguish negative from positive (CD4, CD8, H-2D^S and IL-2R) or low versus high immunofluorescence for CD44. All stained preparations were fixed with 50% ethanol for 30 min on ice.

Cell cycle analyses were carried out on splenic lymphoid populations. Cells were stained with anti-CD4 (clone GK 1.5) or anti-CD8 (clone 2.43), counterstained with mouse anti-rat IgG-FITC (F[ab']$_2$, γ and light chain specificity) (Jackson ImmunoResearch Laboratories, West Grove, PA), fixed with ethanol, incubated with 1 μg/ml of RNase (Sigma Chemical) and then reacted with 50 μg/ml propidium iodide (PI). The cell cycle distribution of tumor cells was determined in a manner similar to that described above for CD4+ and CD8+ T-lymphocytes except that tumor cells were identified by lack of reactivity with anti-H-2D^S and the secondary reagent anti-IgG-PE.

Flow Microfluorometry

All samples were run on a Coulter EPICS 753 flow cytometer. Data were acquired in list mode and analyzed using Becton Dickinson Consort VAX software. Twenty thousand ungated events were collected. CD44 or IL2R expression on CD4+ or CD8+ T-lymphocytes were determined by gating on those cells positive for CD4+ or CD8+ followed by examination of expression of CD25 or CD44 immunofluorescence. When cells were analyzed for cell cycle distribution, PI fluorescence was examined on cells gated for expression of CD4 or CD8 or lack of expression of H-2D^S (tumor cells). Doublet discrimination was applied, and cell cycle distribution was determined by MODFIT (Phoenix Flow Systems, San Diego, CA). Cells in the S+G$_2$M phases of the cell cycle were designated as the proliferative fraction.

Numbers of each cell type in spleen or pooled lymph nodes were computed by multiplying the percentage of the desired cell type by the total number of cells in the lymphoid preparation. Values were determined individually for all animals.

RESULTS

Survival of Cy-Treated and Untreated Tumor-Bearing Mice

SJL mice were injected iv with 10^7 RCS5 tumor cells on day 0 and divided into 2 groups. On day 1, five mice were injected ip with 100 mg/kg of Cy (RCS/Cy mice) while 5 mice remained untreated (RCS5 mice). A second experiment was carried out in which the survival of 4 additional RCS/Cy mice was also monitored. The data from these 2 experiments were pooled in Fig. 1. The median survival time (MST) for RCS5 mice was 8 days, while the MST for RCS/Cy mice was 32 days. Whereas all RCS5 mice were moribund by day 10, the last RCS/Cy mouse was not moribund until day 64. It also is of interest that, whereas RCS5 mice succumbed within one or two days of attaining a high tumor load, some RCS/Cy mice maintained a high tumor load for up to a week prior to succumbing. These data confirm previously published findings that tumor-bearing mice treated with Cy survive

130

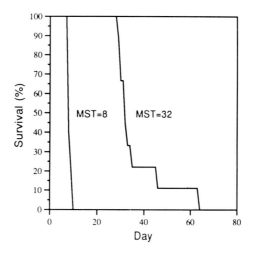

Figure 1. The survival of 9 RCS/Cy and 5 RCS5 mice was followed. The MST is shown for both groups of mice.

much longer than their untreated counterparts (11).

Number and Proliferative Status of Malignant and Nonmalignant Lymphoid Cells of RCS/Cy and RCS5 Mice

A very rapid rate of tumor growth was observed in the spleens of RCS5 mice (Fig. 2). Tumor cells first reached detectable levels (>2% of the lymphoid population) 3 days after injection. By 6 days after injection, nearly 50% of the splenic cellular makeup was replaced by tumor cells. Grossly enlarged spleens contained an average of 2×10^8 tumor cells per spleen. Accompanying this rapid malignant process, corresponding increases in CD4+ and CD8+ T-lymphocytes occurred. Throughout the course of tumor growth, total splenic CD4+ T-lymphocytes considerably exceeded total splenic CD8+ T-lymphocytes reaching a maximum CD4:CD8 ratio of 4.3.

The scenario in RCS/Cy mice was quite different. In contrast to RCS5 mice in which tumor cells were detected by day 3, tumor cells were not detected in the spleens of RCS/Cy mice until day 16 (Fig. 2). However,

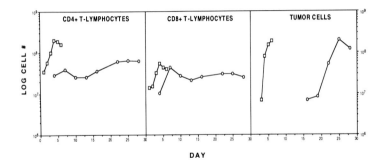

Figure 2. Longitudinal study of mean numbers of CD4+ cells, CD8+ cells and tumor cells per spleen of untreated and Cy-treated tumor-bearing mice. Data are means from 3 mice per group. Study of untreated tumor-bearing mice and Cy-treated tumor-bearing mice was terminated at days 6 and 27, respectively.

by day 19, a rapid increase in numbers of tumor cells commenced leading to a maximum of 2.2×10^8 tumor cells per spleen at day 25. Peaks in total numbers of splenic CD4+ and CD8+ T-lymphocytes were observed at a time (day 7) when tumor cells were undetectable in RCS/Cy mice. At day 7, 40% of the splenic lymphoid cells were CD8+ with a ratio of CD4:CD8 cells of 0.9. After day 7, numbers of CD4+ and CD8+ T-lymphocytes declined in RCS/Cy mice followed by a gradual increase which commenced with the appearance of splenic tumor.

A comparison was also made of the number and percentage of CD4+, CD8+ T-lymphocytes and tumor cells that were proliferating ($S+G_2M$ phases of the cell cycle) in the spleens of RCS5 and RCS/Cy mice (Fig. 3 and 4). Numbers of proliferating CD4+ and CD8+ T-lymphocytes increased steadily in untreated tumor-bearing mice during the rapid course of tumor growth (Fig. 3). Peaks of 3.5×10^7 and 0.9×10^7 proliferating CD4+ and CD8+ T-lymphocytes, respectively, were noted day 6 after the injection of tumor. Also on day 6, the percentages of CD4+ and CD8+ T-lymphocytes in cycle peaked at 22% in conjunction with the number of proliferating tumor cells

132

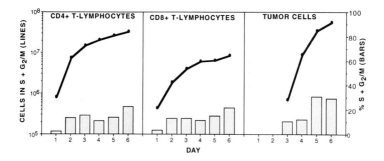

Figure 3. Longitudinal study of mean (3 mice euthanized per day) numbers (lines) and percentages (bars) of CD4+ cells, CD8+ cells and tumor cells that were proliferating (S+G$_2$M phases of the cell cycle) in the spleens of RCS5 mice. Study of RCS5 mice was terminated at day 6.

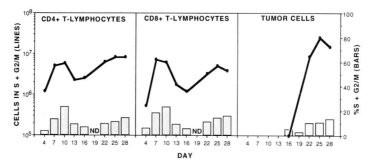

Figure 4. Longitudinal study of mean (3 mice euthanized per day) numbers (lines) and percentages (bars) of CD4+ T-cells, CD8+ T-cells and tumor cells that were proliferating (S+G$_2$M phases of the cell cycle) in the spleens of RCS/Cy mice. Study of RCS/Cy mice was terminated at day 27.

reaching a maximum of 5.9×10^7 per spleen. Proliferating tumor cells were first detected in untreated mice on day 3, from which point a massive increase occurred over the subsequent 3 days. A maximum percentage of 31% tumor cells in cycle was reached on day 5.

The pattern of proliferating nonmalignant and malignant lymphocytes in RCS/Cy mice (Fig. 4) was quite different from that in RCS5 mice. Instead of continuously increasing, the numbers of proliferating splenic CD4+ and CD8+ T-lymphocytes in RCS/Cy mice peaked early (days 7-10) and late (days 25-28) after the injection of tumor cells. The initial peak of proliferative activity occurred when tumor was undetectable, whereas the second peak corresponded with maximization of the number of proliferating tumor cells. The percentage of CD4+ and CD8+ T-lymphocytes that were proliferating peaked at 22-23% at day 10, then declined to 5-6% by day 16, and subsequently rose again to 14-16% by days 25-28 coincident with a heavy tumor load. It is of interest to note, that in RCS/Cy mice the percentage of tumor cells in the $S+G_2M$ phases of the cell cycle never exceeded 13%, whereas a maximum of 31% of tumor cells were proliferating in RCS5 mice.

IL-2R and CD44 Expression of Lymphocytes from RCS/Cy and RCS5 Mice

IL2R and CD44^{hi} expression were assayed as indicators of T-cell activation. The percentage and number of CD4+ T-lymphocytes that were IL2R positive in lymph nodes rose during the rapid course of tumor growth in untreated tumor-bearing mice (Table 1). The percentage of CD4+ T-lymphocytes that were IL2R+ reached a high of 19% 3 days after the injection of tumor cells and maintained an elevated level through day 5. Also, the number of IL2R+ CD4+ T-lymphocytes per lymph node preparation increased markedly from 0.3×10^7 in normal mice to a high of 6.0×10^7 5 days after the injection of tumor cells. Similarly, the percentage of CD44^{hi} expressing CD4+ T-lymphocytes increased from 17% in normal mice to a high of 42% 3 days after the injection of tumor cells. The total number of CD44^{hi} CD4+ T-lymphocytes per lymph node preparation rose from 0.8×10^7 in normal mice to a maximum of 11.0×10^7 in RCS5 mice 5 days after the injection of tumor cells.

In contrast, during the protracted course of tumor growth in RCS/Cy mice, CD4+ T-lymphocytes displayed a pattern of IL2R expression different from that of their

Table 1

IL2R and CD44[hi] Expression by CD4+ T-Lymphocytes
During the Course of Tumor Growth in RCS5 Mice

Day	IL2R+/CD4+[1]		CD44[hi]+/CD4+[1,2]	
	%	No.x10^7	%	No.x10^7
0^3	6.1	0.3	17.0	0.8
1	13.0	1.0	21.0	1.5
2	16.0	1.1	33.0	3.4
3	19.0	2.8	42.0	8.1
4	13.0	2.3	34.0	9.4
5	18.0	6.0	35.0	11.0
6	ND[4]	ND	40.0	8.9

[1] With the exception of day 0, data are means from 3
mice euthanized per day. Lymphocytes were obtained from
selected lymph nodes.
[2] CD44[hi] data is from a second experiment.
[3] Data are means of pooled results from one normal mouse
euthanized per day.
[4] Not done.

RCS5 counterparts (Table 2). The percentage of CD4+ T-
lymphocytes expressing IL2R increased from 5.9% in
normal mice to 17% on day 7 after the injection of
tumor. However, instead of continuing to increase, the
percentage of IL2R+ CD4+ T-lymphocytes fell back to 10-
11%, a level which was maintained until day 19 when a
marked increase commenced. For the most part, CD44[hi]
expressing CD4+ T-lymphocytes continually increased in
percentage and number during the protracted course of
tumor growth. The contrast between the biphasic IL2R
expression and the continuous increase in CD44[hi] expression on CD4+ T-lymphocytes from Cy-treated tumor-bearing
mice may be explained by the known transient nature of
IL2R expression compared with the stable display of the
CD44[hi] phenotype once it is acquired (12,17).

Activation markers also were compared on CD8+ T-
lymphocytes from RCS5 and RCS/Cy mice. In RCS5 mice,

Table 2

IL2R and CD44[hi] Expression by CD4+ T-Lymphocytes
During the Course of Tumor Growth in RCS/Cy Mice

Day	IL2R+/CD4+[1]		CD44[hi]+/CD4+[1,2]	
	%	No.x10[7]	%	No.x10[7]
0[3]	5.9	0.3	13.0	0.7
4	13.0	0.4	20.0	0.8
7	17.0	0.5	26.0	1.3
10	10.0	0.4	28.0	1.3
13	10.0	0.3	30.0	1.7
16	11.0	0.5	35.0	1.6
19	18.0	0.7	52.0	3.8
25	22.0	2.0	57.0	2.0
28	37.0	3.4	43.0	3.7

[1] With the exception of day 0, data are means from 3 mice euthanized per day. Lymphocytes were obtained from selected lymph nodes.
[2] CD44[hi] data is from a second experiment.
[3] Data are means of pooled results from one normal mouse euthanized per day.

the percentage of CD8+ T-lymphocytes that expressed IL2R or CD44[hi] increased continuously, reaching highs of 27% on day 5 and 48% on day 6, respectively (Table 3). Numbers of IL2R+ CD8+ and CD44[hi]+ CD8+ lymphocytes also generally increased coincident with the rapid rate of tumor growth in untreated mice.

Again, a different situation was noted during the protracted course of tumor growth in RCS/Cy mice. The percentage of CD8+ T-lymphocytes that were IL2R+ increased from 4.4% in normal mice (day 0) to 14% at day 7 (Table 4). However, instead of continuing to increase in percentage, the frequency of CD8+ cells that were IL2R+ declined to 10% by day 10 and generally remained at this level through day 19. Thereafter, marked increases in the percentage of IL2R+ CD8+ T-lymphocytes were noted, reaching 42% at day 28 when a significant

Table 3

IL2R and CD44[hi] Expression by CD8+ T-Lymphocytes During the Course of Tumor Growth in RCS5 Mice

Day	IL2R+/CD8+[1]		CD44[hi]+/CD8+[1,2]	
	%	No.x10^7	%	No.x10^7
0[3]	3.8	0.1	12.0	0.3
1	9.3	0.4	13.0	0.5
2	14.0	0.4	29.0	1.2
3	20.0	0.9	37.0	1.9
4	20.0	0.8	37.0	1.6
5	27.0	1.4	46.0	2.6
6	ND[4]	ND	48.0	1.3

[1] With the exception of day 0, data are means from 3 mice euthanized per day. Lymphocytes were obtained from selected lymph nodes.
[2] CD44[hi] data is from a second experiment.
[3] Data are means of pooled results from one normal mouse euthanized per day.
[4] Not done.

tumor load was evident. This biphasic rise in percentage of activated CD8+ T-lymphocytes was also noted for CD44[hi]+ expressing CD8+ T-lymphocytes, with an initial peak of 45% on day 7 and a second peak on day 19. Peaks in numbers of CD44[hi]+ CD8 T-lymphocytes were also observed on days 7 and 19.

DISCUSSION

The efficacy of the RCS/Cy treatment is derived from the antitumor effects of Cy, from the inhibitory effect of Cy on activated CD4+ T-lymphocytes, and from the appearance of a population of CD8+ T-lymphocytes able to suppress the proliferation of CD4+ T-lymphocytes in response to tumor cells (11). The data reported herein, confirmed the effectiveness of this treatment by demonstrating markedly increased survival, a significant delay in the appearance of tumor cells and a lower

Table 4

IL2R and CD44[hi] Expression by CD8+ T-Lymphocytes
During the Course of Tumor Growth in RCS/Cy Mice

| Day | IL2R+/CD8+[1] | | CD44[hi]+/CD8+[1,2] | |
	%	No.x10[7]	%	No.x10[7]
0[3]	4.4	0.1	8.0	0.2
4	13.0	0.3	20.0	0.6
7	14.0	0.4	45.0	2.1
10	10.0	0.4	34.0	1.1
13	8.0	0.2	30.0	1.0
16	11.0	0.3	35.0	1.0
19	11.0	0.2	61.0	2.5
25	21.0	0.7	51.0	0.6
28	42.0	1.0	44.0	1.6

[1] With the exception of day 0, data are means from 3
mice euthanized per day. Lymphocytes were obtained from
selected lymph nodes.
[2] CD44[hi] data is from a second experiment.
[3] Data are means of pooled results from one normal mouse
euthanized per day.

fraction of tumor cells in cycle in RCS/Cy than in RCS5
mice. The decreased capacity of CD4+ T-lymphocytes from
RCS/Cy mice to support tumor growth has been shown to be
the principal component responsible for this treatment's
success (11). Therefore, data as to the longitudinal
interrelationships between levels of CD4+ T-lymphocytes,
CD8+ T-lymphocytes and tumor cells was expected to provide further insight as to the mechanisms operative in
this treatment.

Three predictions formed the premise for this
study, and each prediction was supported by the results.
First, the data indicated that _in situ_ activation/proli-
feration of CD4+ T-lymphocytes was inferior in RCS/Cy
compared with RCS5 mice. In RCS5 mice "explosive" tumor
growth was characterized by a rapid increase in number
and percentage of activated (IL2R+ and CD44[hi]+) and

proliferating (S+G$_2$M phases of the cell cycle) CD4+ T-lymphocytes. In contrast, the protracted course of tumor growth in RCS/Cy mice was characterized by early (days 7-10) and late (days 25-28) peaks of activated/proliferating CD4+ T-lymphocytes with an intervening period of depressed activation/proliferation. The late peak of activation/proliferation was probably analogous to the increase seen in RCS5 mice since, in both instances, CD4+ T-lymphocyte activation/proliferation occurred coincident with the escalating presence of tumor cells. Such an increase in CD4+ cell activity was to be expected given the fact that SJL tumor cells stimulate proliferation of syngeneic CD4+ T-lymphocytes both in vitro and in vivo (18). However, the early (days 7-10) peak of CD4+ T-lymphocyte activation/proliferation in RCS/Cy mice was surprising given the fact that tumor cells were undetectable at this point, and, indeed, tumor cells were not detected until day 16. It is conceivable that the early peak of CD4+ T-lymphocyte activation/proliferation was in response to minimal numbers of tumor cells. Indeed, increased activation/proliferation of CD4+ T-lymphocytes, was noted in untreated tumor-bearing mice 2 days before tumor cells were detected by the relatively insensitive flow microfluorometric procedure used. However, of importance, instead of a subsequent steady increase in activated/proliferating CD4+ T-lymphocytes, as is noted in RCS5 mice, a period ensued in RCS/Cy mice in which there was a decline in CD4+ T-lymphocyte activation/proliferation. It would appear that this transient rise in CD4+ T-lymphocyte activation/proliferation was an abortive attempt of occult tumor to stimulate CD4+ T-lymphocyte activation/proliferation in order to generate a microenvironment conducive to maximum proliferation of tumor cells. Since CD8+ T-lymphocytes isolated between days 7 and 13 from RCS/Cy mice suppress in vitro CD4+ T-lympho-

cyte proliferation in response to SJL tumor cells
(MLTI), it is most likely that this population of CD8+
T-lymphocytes was responsible for controlling the tumor-
driven in situ activation/proliferation of CD4+ T-lymphocytes in RCS/Cy mice. Therefore, an environment was
created nonconducive to tumor cell proliferation (11).

Based on the markedly increased survival of RCS/Cy
compared with RCS5 mice, the second goal of this study
was to test the prediction that proliferation of tumor
cells in RCS/Cy mice would be inferior to that in RCS5
mice. The results supported this prediction. Tumor
cells were not detected until day 16 in RCS/Cy mice.
This contrasts with detection of tumor cells only 3 days
after their injection in RCS5 mice. Furthermore, whereas a maximum of 31% of the tumor cells were proliferating in RCS5 mice, the maximum percentage of proliferating tumor cells in RCS/Cy mice was only 13%. The best
explanation for these observations was that, in RCS/Cy
mice, tumor cells were deprived of sufficient contact
with activated CD4+ T-lymphocytes and/or CD4+ T-lympho-
cyte-derived products to promote maximum tumor cell
proliferation. Growth of SJL lymphomas is known to be
inferior in environments deficient in either syngeneic
CD4+ T-lymphocytes or products derived there from
(6,9,10).

The final goal of this study was to gain further
insight as to the identity of the population of CD8+ T-
lymphocytes responsible for suppressing CD4+ T-lymphocyte proliferation in response to tumor cells. Since it
is known that CD8+ T-lymphocytes derived from RCS/Cy
mice 7-13 days after the injection of tumor cells possess the capacity to specifically suppress the MLTI
(11), it was not surprising to find that a peak in
numbers and percentages of activated/proliferating CD8+
T-lymphocytes occurred within this time frame. One may
speculate that the suppressive fraction of CD8+ T-lym-

phocytes was contained within this population of acti-
vated/proliferating CD8+ T-lymphocytes. To prove this
point, it will be necessary to enrich for this cell
fraction and demonstrate markedly enhanced suppressive
capacity.

From this study it can be seen that the ratio of
total, activated or proliferating CD4+ to CD8+ T-lymphocytes was predictive of the suitability of an in vivo
environment for proliferation of SJL tumor cells. The
greater the excess of CD4+ to CD8+ cells, the more
conducive was the environment to growth of the RCS5 SJL
lymphoma. During the maximum phase of tumor growth in
untreated mice, CD4+ T-lymphocytes exceeded CD8+ T-lymphocytes by greater than 3 fold. In contrast, during
the period prior to the appearance of detectable tumor
cells in RCS/Cy mice (earlier than day 16), the ratio of
CD4+ to CD8+ cells never exceeded 2 and in some instances was less than 1. Therefore, a shift of the balance
of CD4+ to CD8+ T-lymphocytes more in favor of a suppressive CD8+ population led to conditions nonconducive
to CD4+ T-lymphocyte and, in turn, tumor cell proliferation.

In response to an antigenic stimulus, such as
provided by syngeneic SJL tumor cells, CD4+ T-lymphocytes undergo a cascade of events which culminates in
proliferation (19). The data reported herein, indicate
that the CD8+ cell-rich environment of Cy-treated mice
sets the stage for deficient activation and proliferation of CD4+ T-lymphocytes in response to tumor cells.
It will be of interest to determine how early in the
activation cascade CD8+ T-lymphocytes from Cy-treated
tumor-bearing mice suppress the response of CD4+ T-
lymphocytes to tumor cells, thereby inhibiting growth of
this CD4+ cell-dependent murine B-cell lymphoma. It
will also be important to determine the mechanism by
which the RCS/Cy treatment shifts the lymphocytic bal-

ance in favor of suppressive CD8+ T-lymphocytes. Furthermore, considerable insight will be obtained through knowledge of the mechanism by which CD8+ T-lymphocytes mediate their down regulation of tumor-stimulated CD4+ T-lymphocyte proliferation so critical to the growth of these lymphomas.

ACKNOWLEDGEMENTS

Supported by USPHS Grant CA52603. The Department of Immunology and Microbiology Flow Cytometry Facility is supported, in part, by the Center for Molecular Biology of Wayne State University. We thank Mr. Mark KuKuruga for his advice.

REFERENCES

1. Murphy ED: SJL/J, a new inbred strain of mouse with a high early incidence of reticulum cell neoplasms. Proc. AACR 4:46, 1963.
2. Wanebo HJ, Gallmeir WM, Boyse EA, Old LJ: Paraproteinemia and reticulum cell sarcoma in an inbred mouse strain. Science 154:901-903, 1966.
3. Pattengale PK, Taylor CR: Experimental models of lymphoproliferative disease. The mouse as a model for human non-Hodgkin's lymphomas and related leukemias. Am. J. Pathol. 113:237-265, 1983.
4. Rosenberg SA: Karnofsky memorial lecture. The low-grade non-Hodgkin's lymphomas: Challenges and opportunities. J. Clin. Oncol. 3:299-310, 1985.
5. Ponzio NM, Brown PH, Thorbecke GJ: Host-tumor interactions in the SJL lymphoma model. Int. Rev. Immunol. 1:273-301, 1986.
6. Lerman SP, Carswell EA, Chapman J, Thorbecke GJ: Properties of reticulum cell sarcomas in SJL/J mice. III. Promotion of tumor growth in irradiated mice by normal lymphoid cells. Cell. Immunol. 23: 53-67, 1976.
7. Lerman SP, Chapman-Alexander J, Umetsu D, Thorbecke GJ: Properties of reticulum cell sarcomas in SJL/J mice. VII. Nature of normal lymphoid cells proliferating in response to tumor cells. Cell. Immunol. 43:209-213, 1979.
8. Lasky JL, Thorbecke GJ: Characterization and growth factor requirements of SJL lymphomas. II. Interleukin-5 dependence of the in vitro cell line cRCS-X, and influence of other cytokines. Eur. J. Immunol. 19:365-371, 1989.

9. Ohnishi K, Bonavida B: Regulation of Ia+ reticulum cell sarcoma (RCS) growth in syngeneic SJL/J mice. I. Inhibition of tumor growth by passive administration of L3T4 monoclonal antibody before or after tumor inoculation. J. Immunol. 138:4524-4529, 1987.

10. Aliasaukas RM, Ponzio NM: T-helper cell specific monoclonal antibody inhibits growth of B-cell lymphomas in syngeneic SJL/J mice. Cell. Immunol. 119:286-303, 1989.

11. Thrush GR, Placey JL, Valeriote FA, Lerman SP: The CD4 cell dependency of SJL/J B-cell lymphomas as a target for cyclophosphamide therapy. Cancer Comm. 1:301-310, 1989.

12. Smith KA: Interleukin-2: Inception, impact, and implications. Science 240:1169-1176, 1988.

13. Butterfield K, Fathman CG, Budd RC: A subset of memory CD4+ helper T lympocytes indentified by expression of Pgp-1. J. Exp. Med. 169:1461-1466, 1989.

14. MacDonald HR, Budd RC, Cerottini JC: Pgp-1 (Ly 24) as a marker of murine memory T lymphocytes. Current Topics Microbiol. Immunol. 159:97-109, 1990.

15. Sopchak L, King SR, Miller DA, et al: Progression of transplanted SJL/J lymphomas attributed to a single aggressive H-2D^S negative lymphoma. Cancer Res. 49:665-671, 1989.

16. Rosloniec EF, Kuhn MH, Genyea CA, et al: Aggressiveness of SJL/J lymphomas correlates with absence of H-2D^S antigens. J. Immunol. 132:945-952, 1984.

17. Budd RC, Cerottini J-C, Horvath C, et al: Distinction of virgin and memory T lymphocytes. Stable acquisition of the Pgp-1 glycoprotein concomitant with antigenic stimulation. J. Immunol. 138:3120-3129, 1987.

18. Ponzio NM, Lerman SP, Chapman JM, Thorbecke GJ: Properties of reticulum cell sarcomas in SJL/J mice. IV. Minimal development of cytotoxic cells despite marked proliferation to syngeneic RCS in vivo and in vitro. Cell. Immunol. 32:10-22, 1977.

19. Roa A: Signaling mechanisms in T cells. Critical Reviews Immunol. 10:495-519, 1991.

10

GENE MAPPING AND MOLECULAR CYTOGENETICS USING FLOW-
SORTED CHROMOSOMES

N.P. Carter

BIVARIATE ANALYSIS OF HUMAN CHROMOSOMES

The bivariate analysis of human chromosome suspensions by flow cytometry (1) enables the majority of the
chromosome types to be resolved and sorted. Chromosomes
are stained with two DNA-specific fluorochromes, Hoechst
33258 which has specificity for AT-rich DNA and Chromomycin A3 with specificity for GC-rich DNA. In the flow
cytometer, the chromosomes are passed sequentially
through two spatially separated laser beams, the first
operated using the UV lines (351-364 nm) to excite
Hoechst fluorescence and the second operated at 457.9 nm
to excite Chromomycin fluorescence. The intensity of
the fluorescence emissions from the two dyes are measured independently and correlated on the bivariate flow
karyotype where chromosome types are resolved by DNA
content and base pair ratio (see Fig. 1). All of the
chromosome types can be resolved in this way with the
exception of chromosomes 9-12 which display similar DNA
content and base pair ratio. This degree of separation,
which is achievable on unmodified, commercially available flow sorters (2), enables fractions, highly enriched for a particular chromosome type, to be produced
for use in the production of DNA libraries, for gene
mapping and for use in cytogenetic analysis of metaphase
chromosomes and interphase nuclei.

144

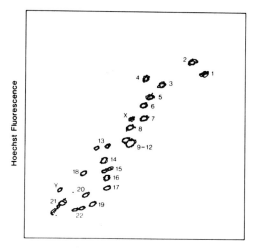

Chromomycin A3 Fluorescence

Figure 1. Bivariate flow karyotype of chromosomes from lymphoblastoid cells from a normal male (reprinted from ref. 3).

GENE MAPPING

One of the current major aims of human molecular genetics is the location of disease-associated genes at a molecular level on the DNA strand. Subsequent sequencing of the entire gene leads to improved prenatal diagnosis, improved family genetic counselling, improved disease management and the potential for gene therapy. As structural genes are estimated to represent less than 17% (4) of the human genome, the location of disease-associated genes is a daunting task. A common approach is to identify DNA markers which show genetic linkage to the disease in affected families. These markers can be assigned to a particular chromosome and placed into a physical map which comprises an ordered series of both expressed and anonymous DNA sequences localized along the chromosome. On the physical map of the chromosome, the markers most closely flanking the disease locus can be identified. The more detailed the physical map becomes with an increasing number of ordered sequences,

the smaller becomes the distance between the disease
locus and the flanking markers until the disease gene
itself can be localized. It is clear that an accurate,
detailed, physical map of each chromosome is vital to
the localization of disease genes in this way.

Flow-sorted chromosomes are a valuable resource for
the initial chromosome localization of DNA sequences and
for the construction of physical maps. We have used two
approaches, the use of flow-sorted chromosome dot blots
and the specific amplification of DNA sequences from
small numbers of flow-sorted chromosomes using the
polymerase chain reaction (PCR).

Chromosomes flow-sorted onto nitrocellulose filters
can be used as targets for the direct hybridization of
cloned DNA sequences (5). These flow-sorted dot blots
can be used to provide an initial chromosomal localization for a DNA sequence. They are produced by sorting
10,000 of each chromosome type as separate dots onto
nitrocellulose filter discs, two chromosome types per
disc. Spreading of the sorted chromosomes on the disc
is restricted during sorting by the application of mild
aspiration from below. The dots of chromosomal DNA are
denatured by alkali treatment, neutralized and then
baked onto the disc. A panel of 11 discs comprises all
of the different chromosome types (chromosome 9-12 being
sorted as a single dot). The DNA sequence to be localized is radiolabelled and then hybridized to the panel of
filters. Subsequent autoradiography then reveals the
spot of DNA to which the probe has hybridized and thus
identifies the chromosomal location of the DNA sequence
(see Fig. 2). Many sequences have been localized to
specific chromosomes in this way (6-11). We have found
that this approach works efficiently only for highly
purified probes which neither contain repetitive sequences nor are associated with pseudogenes; however,
flow-sorted dot blots have the advantage that sequence

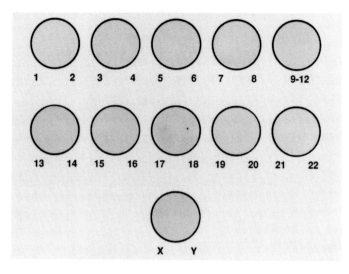

Figure 2. Localization of the gene for human ferrochelatase to chromosome 18 by flow-sorted dot blot analysis
(reprinted from ref. 11).

data is not required.

An alternate approach for the chromosomal localization of DNA sequences is to use enzymatic amplification of the sequence from small numbers of chromosomes
sorted directly into PCR tubes (12,13). We have found
200 chromosomes sufficient as target DNA for most specific primer pairs. Chromosomes of each type are sorted
directly into separate PCR tubes containing 30 μl of
purified water to produce a panel of 21 tubes representing the whole genome (chromosomes 9-12 sorted together).
PCR buffer, dNTPs, specific primer pairs and Taq polymerase are added to each tube to make a 50 μl reaction,
overlaid with mineral oil and then amplified typically
for 40 cycles. An aliquot of each tube (10-20 μl) is
run on an agarose gel and stained with Ethidium Bromide.
Specific product of the correct molecular weight is then
detected in one of the lanes using UV illumination thus
localizing the sequence to the appropriate chromosome
(see Fig. 3). In practice, to reduce the number of PCRs

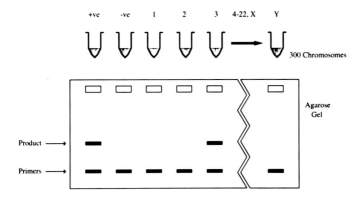

Figure 3. Schematic representation of gene mapping by PCR amplification on sorted chromosomes.

required for a localization, we have adopted a two-step analysis. In the initial analysis, pools of 200 of each of 4 chromosome types (i.e. 1-4, 5-8, 9-12, 13-16, 17-20 and 21, 22, X and Y) are used as target DNA for amplification. The PCR is performed, the reaction mixtures are run on an agarose gel and the tube in which product has been generated is identified. The same primers are then used with tubes containing 200 of the individual chromosomes within the pool and the reaction mixtures run on a second agarose gel. The tube in which product is detected maps the sequence to that particular chromosome (see Fig. 4). In this way, only 10 PCR reactions are required for chromosomal localization compared with the 21 reactions required if chromosome pools are not used.

The precision of mapping sequences, both with dot blots and with the PCR approach, can be refined by the use of cell lines derived from individuals with balanced translocations. In most cases, the result of the translocation is the production of two aberrant chromosomes which are different in size from the normal chromosomes from which they are derived. Usually, these derivative

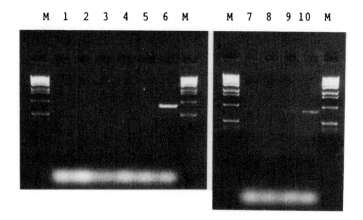

Figure 4. Agarose gel electrophoresis of the products of PCR amplification of the human SRY gene on sorted chromosomes. (M-molecular weight markers; Lane 1-chromosome pool 1-4; Lane 2-chromosome pool 5-8; Lane 3-chromosome pool 9-12; Lane 4-chromosome pool 13-16; Lane 5-chromosome pool 17-20; Lane 6-chromosomes 21, 22, X and Y; Lane 7-chromosome 21; Lane 8-chromosome 22; Lane 9-chromosome X; Lane 10-chromosome Y).

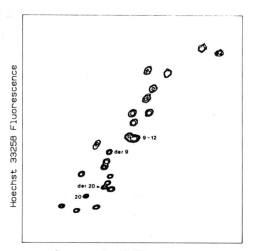

Figure 5. Flow karyotype of a lymphoblastoid cell line derived from an individual with a translocation between chromosomes 9 and 20. The normal and derivative chromosomes involved in the translocation are indicated.

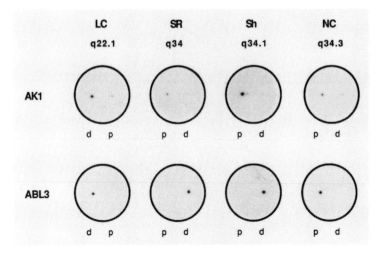

Figure 6. Hybridization of two DNA probes to translocation chromosome flow-sorted dot blots. The chromosome
dot representing the proximal part and distal part of
chromosome 9 for each translocation are indicated by p
and d, respectively (reprinted from ref. 14).

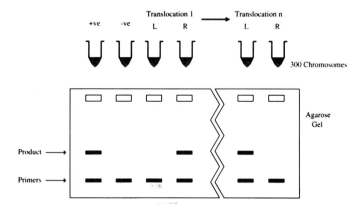

Figure 7. Schematic representation of gene mapping
using PCR amplification of chromosome translocation
products.

chromosomes can be resolved on the bivariate flow karyotype (see Fig. 5) and so sorted directly onto nitrocellulose filter discs or into PCR tubes. A probe or
sequence that has been mapped previously to one of the
chromosomes involved in the translocation can then be
used with the two derivative chromosomes to determine
whether the sequence maps proximal to or distal to the
translocation breakpoint (Fig. 6). We have used this
approach extensively with both flow-sorted dot blots and
with PCR amplification (Fig. 7) to create a physical map
of the long arm of chromosome 9. We have collected over
20 cell lines with different breakpoints on the long arm
of chromosome 9 and thus can assign any probe or sequence into one of the 21 intervals (Fig. 8). The
resolution of the physical map produced in this way is
limited only by the number of breakpoints and the physical distance between them.

MOLECULAR CYTOGENETICS

Flow-sorted chromosomes can be extremely helpful in
the cytogenetic analysis of abnormal karyotypes. The
production of libraries from large numbers of flow-
sorted chromosomes has led to their use for chromosome
painting of metaphase spreads using fluorescence _in situ_
hybridization (FISH). DNA from the library is labelled
typically with a hapten (e.g. biotin) using nick-trans-
lation with modified di-nucleotide triphosphates.
Hybridization of the library to metaphase spreads on
glass slides is then visualized using a fluorescent
detection system (e.g. avidin-FITC) and direct observation on the fluorescence microscope (15). For analysis
of metaphases from an individual with an abnormal karyotype or from a tumour, appropriate chromosome-specific
libraries, as indicated by routine banding analysis, are
used to highlight the aberration and the chromosome
breakpoints involved (16). While chromosome painting

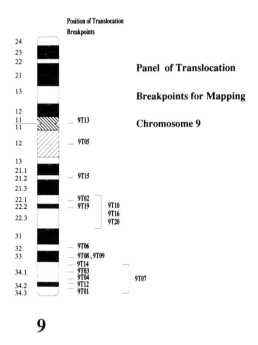

Figure 8. Panel of translocation breakpoints for mapping the long arm of chromosome 9.

with libraries works well for cases where chromosome banding can identify the chromosomes involved in the rearrangement and thus indicate which chromosome libraries to use, painting with chromosome libraries cannot be used to characterize either interstitial deletions or the precise origin of <u>de novo</u> unbalanced rearrangements. For such cases, we have developed a technique we have called reverse chromosome painting where the chromosome paint is generated rapidly by PCR from small numbers of the aberrant chromosome itself, flow-sorted from either blood cultures or cell lines of the affected individual, and the PCR products painted back onto normal metaphase spreads. In this way, the origin of aberrant chromosomes can be visualized directly from the hybridization pattern on normal chromosomes (17). Crucial for the reverse chromosome painting technique is the ability to amplify in a general and representative way DNA from small numbers of flow-sorted chromosomes. For this purpose, we have developed a novel, partially degenerate, PCR primer for use in a dual annealing temperature PCR protocol which we have called Degenerate Oligonucleotide-Primed PCR (DOP-PCR). The primer, 6MW (5' CCG ACT CGA GNN NNN NAT GTG G 3', where N = any base), is designed with 6 specific 3' bases and 10 specific 5' bases separated by a degenerate region of 6 bases (3). The target DNA (300 chromosomes) is subjected to 9 low temperature annealing cycles (94°C for 1 min, 30°C for 1 min, transition to 72°C at 0.23°C-/sec, 72°C for 2 mins) followed immediately by 30 high annealing temperature cycles (94°C for 1 min, 62°C for 1 min, 72°C for 1 min). During the low annealing temperature cycles, we believe that priming is restricted by the partial degeneracy to chiefly the 6 most 3' bases resulting in priming sites on average every 2 kb (18). In this way, initial product representative of the entire chromosome is generated, tailed at one end with the primer sequence and at the other with the sequence complementary to the primer. During the high annealing temperature cycles, both the 3' and the 5' bases are required for stable priming so that only product tailed during the initial cycles is amplified further. Biotinylation of the PCR product for chromosome painting is achieved simply by a second round of high annealing temperature cycles incorporating biotin-11-dUTP in the reaction mix.

We have been applying this reverse chromosome painting technique for the analysis of routine cases submitted for diagnosis to the East Anglian Regional Cytogenetics Laboratory. Fig. 9 shows the flow karyotype obtained from a mother referred for genetic analysis after two fetal losses due to neural tube defects. Chromosome banding analysis indicated an insertion from

the long arm of chromosome 13 into the short arm of
chromosome 1 and while the deleted chromosome 13 is
resolved clearly on the flow karyotype, the chromosome 1
with the insertion is not resolved from the normal
chromosome 1. Reverse painting DOP-PCR products from
the combined chromosome 1 peak onto normal chromosomes
is shown in Fig. 10a. In addition to the two normal
chromosomes 1, signal was detected from distal 13q31 to
distal 13q33, clearly demonstrating the precise origin
of the insertion into the aberrant chromosome 1. The
involvement of the same region also could be demonstrated from the reverse painting of the deleted chromosome
13 onto normal metaphase spreads as shown in Fig. 10b.
In this case, the origin of the insertion is demonstrated by the segment of the normal chromosome 13 not painted. While the origin of the insertion is demonstrated
clearly, the position of the insertion into chromosome 1
cannot be visualized by the reverse painting technique.

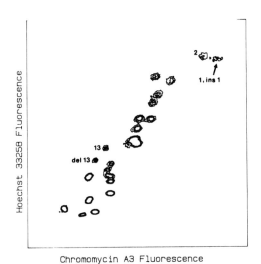

Figure 9. Flow karyotype of a patient with an inserti-
onal translocation between chromosomes 13 and 1. The
normal and derivative chromosomes involved in the translocation are indicated (reprinted from ref. 17).

154

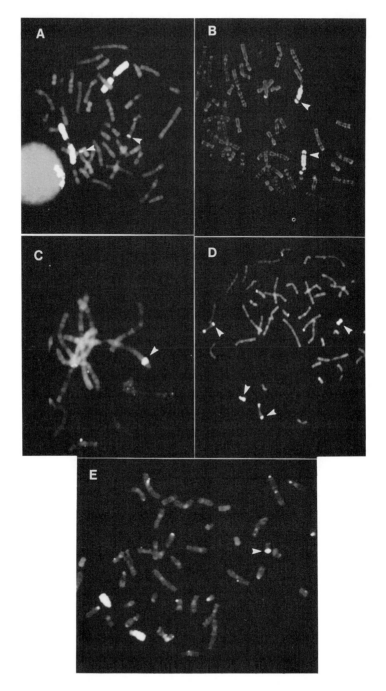

Figure 10. Analysis of abnormal chromosomes using reverse and normal chromosome painting. See text for details of each panel (reprinted from ref. 17).

However, conventional painting of the patient's chromosomes with a normal chromosome 13 paint (DOP-PCR product) allows us to visualize the insertion and determine the chromosome breakpoints (Fig. 10c).

We have found the reverse chromosome painting technique particularly powerful for the analysis of <u>de novo</u> chromosomal aberrations. Fig. 11 shows the flow karyotype obtained from a 1-year-old boy with hypotonia and delayed motor development whose GTL banded karyotype showed additional material on the short arm of one chromosome 21. This 21p+ chromosome is resolved from the normal chromosome 21 in the flow karyotype and thus both chromosomes could be sorted separately for reverse chromosome painting. The DOP-PCR product from the normal chromosome 21 from the patient painted onto normal chromosomes showed hybridization only to chromosomes 21. However, the derivative chromosome 21 paint showed hybridization on normal metaphase spreads to chromosomes 21, acrocentric chromosome short arms and to

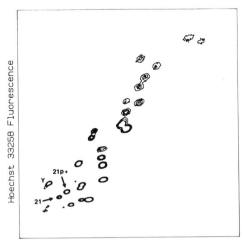

Chromomycin A3 Fluorescence

Figure 11. Flow karyotype of a child with additional material on one chromosome 21. The normal and abnormal chromosomes 21 are indicated (reprinted from ref. 17).

the region q32.1——▸qter of chromosome 14 (Fig. 10d).
Based on this analysis, the child has a duplication and
partial trisomy of chromosome 14 involving region
14q32.1——▸qter. The diagnosis of chromosome 14 as the
source of the duplication was confirmed by applying a
normal chromosome 14 DOP-PCR paint onto the patient's
chromosomes. In addition to the two chromosomes 14,
signal was detected on the short arm of one chromosome
21 (Fig. 10e). While this conventional painting analysis was able to confirm the chromosome 14 origin of the
duplication, only reverse chromosome painting was able
to demonstrate the region of chromosome 14 involved.

CONCLUSION

In this chapter, I have attempted to demonstrate
the usefulness of flow-sorted chromosomes in gene mapping and molecular cytogenetics. For the next few
years, the construction of increasingly refined physical
maps of each chromosome will be an imperative for the
various Human Genome Projects in progress around the
world. Flow-sorting of naturally occurring translocation products and the mapping of cloned sequences using
these chromosomes will continue to have a useful role in
the construction of physical maps. Such mapping is
facilitated greatly by the ability to design primer
pairs from published sequences for subsequent PCR amplification with small numbers of flow-sorted chromosomes.

I have also shown how DOP-PCR of small numbers of
flow-sorted chromosomes can be used for the rapid analysis of abnormal chromosomes. This reverse painting
technique has particular application to the analysis of
tumour cell lines where multiple rearrangements make
conventional banding analysis difficult and inaccurate.
The ability to amplify sorted chromosomes in a general
and representative way opens up possibilities for direct
cloning as well as coincidence, deletion and breakpoint

cloning. Using these techniques, it should be possible
to saturate rapidly small chromosomal regions with new
markers to facilitate the localization of disease associated genes.

ACKNOWLEDGEMENTS

The work described here was carried out with support from the Medical Research Council of the UK, the UK
Human Genome Project, and the Cancer Research Campaign.
I would particularly like to acknowledge Professor M.A.
Ferguson-Smith; H. Telenius who designed the DOP-PCR
primers and protocol; A.H. Pelmear, M.T. Perryman, L.R.
Willatt and M.A. Leversha for FISH analyses; the staff
of the East Anglian Regional Genetics Service for refer-
ral and cytogenetic analysis of patients. I am grateful
to G. van den Engh for the program ANALIST used to
produce flow karyotype contour plots.

REFERENCES

1. Gray JW, Langlois RG, Carrano AV, et al: High
   resolution chromosome analysis: One and two parameter flow cytometry. Chromosoma 73:9-27, 1979.
2. Carter NP, Ferguson-Smith ME, Affara NA, et al:
   Study of chromosome abnormality in XX males using
   bivariate flow karyotype analysis and flow sorted
   dot blots. Cytometry 11:202-207, 1990.
3. Telenius H, Pelmear AH, Tunnacliffe A, et al:
   Cytogenetic analysis by chromosome painting using
   degenerate oligonucleotide-primed-polymerase chain
   reaction amplified flow-sorted chromosomes. Genes,
   Chromosomes Cancer 4:257-263, 1992.
4. Connor JM, Ferguson-Smith MA: Essential Medical
   Genetics. Blackwell Scientific Publications, 1987.
5. Bernheim A, Metezeau P, Guellaen G, et al: Direct
   hybridization of sorted human chromosomes: Localization of the Y chromosome on the flow karyotype.
   Proc. Natl. Acad. Sci. USA 80:7571-7575, 1983.
6. Lebo RV, Gorin F, Fletterick RJ, et al: High-resolution chromosome sorting and DNA spot-blot analysis assign McArdle's syndrome to Chromosome 11.
   Science 225:57-59, 1984.
7. Collard JG, de Boer PAJ, Janssen JWG, et al: Gene
   mapping by chromosome spot hybridization. Cytometry 6:179-185, 1985.

8. Lebo RV, Bruce BD: Gene mapping with sorted chromosomes. Methods Enzymol. 151:292-313, 1987.
9. Arai N, Nomura D, Villaret D, et al: Complete nucleotide sequence of the chromosomal gene for human IL-4 and its expression. J. Immunol. 142: 274-282, 1989.
10. Neckelmann N, Warner CK, Chung A, et al: The human ATP synthase ß subunit gene: Sequence analysis, chromosomal assignment and differential expression. Genomics 5:829-843, 1989.
11. Whitcome DM, Carter NP, Albertson DG, et al: Assignment of the human ferrochelatase gene (FECH) and a locus for protoporphyria to chromosomes 18q22. Genomics 11:1152-1154, 1991.
12. Schmitz A, Olschwang S, Chaput B, et al: Oncogene detection by enzymatic amplification on flow sorted chromosomes. Nucleic Acids Res. 17:816, 1989.
13. Cotter F, Nasipuri S, Lam G, Young BD: Gene mapping by enzymatic amplification from flow-sorted chromosomes. Genomics 5:470-474, 1989.
14. Carter NP, Hampson RM, Harris RM, et al: Gene mapping using flow-sorted chromosomes. Proc. Royal Microscopical Soc. 1:511-514, 1990.
15. Pinkel D, Landegent J, Collins C, et al: Fluorescence in situ hybridisation with human chromosome-specific libraries: Detection of trisomy 21 and translocations of chromosome 4. Proc. Natl. Acad. Sci. USA 85:9138-9142, 1988.
16. Hulten MA, Gould CP, Goldman ASH, Waters JJ: Chromosome in situ suppression hybridisation in clinical cytogenetics. J. Med. Genet. 28:577-582, 1991.
17. Carter NP, Ferguson-Smith MA, Perryman MT, et al: Reverse chromosome painting: A method for the rapid analysis of aberrant chromosomes in clinical cytogenetics. J. Med. Genet. 29:299-307, 1992.
18. Telenius H, Carter NP, Bebb CE, et al: Degenerate oligonucleotide-primed PCR (DOP-PCR): General amplification of target DNA by a single degenerate primer. Genomics 13:718-725, 1992.

11

THE CLINICAL POTENTIAL OF DNA CONTENT PARAMETERS IN
HUMAN PEDIATRIC AND ADULT SOLID TUMORS

Mark Zalupski and John F. Ensley

INTRODUCTION

The ability to predict the natural history and
treatment outcome of human cancers, particularly for an
individual patient, has long been a goal of cancer
researchers. Knowledge concerning the natural history
of a particular cancer was acquired by careful clinical
observation long before effective treatments were developed and eventually evolved into our current clinical
staging systems. Over the last century, light microscopic and tissue staining technologies have evolved
which have allowed the careful correlation of pathological features of tumors and the clinical outcome of the
patients that harbor them. Until very recently, individual patient treatment decisions and the design of
clinical trials were based primarily on these two types
of parameters; clinical stage and histopathology.

Cytotoxic therapy, first ionizing radiation and
more recently cytotoxic drugs, have uncovered additional
differences in the underlying biology of human cancers
characterized by response and recurrence patterns.
Correlations between stage and histopathology were first
made with these different response groups, and in some
tumors were found to be prognostically important. The
ability to predict clinical and treatment outcome in
patients with cancer remains critical in that it would
allow the rational design of adjuvant clinical trials

through the identification of patients at risk for
recurrence following local therapy, the identification
of patients likely to benefit from organ preservation
strategies, the design of cytotoxic regimen intensity
consistent with the degree of tumor response, and the
study of molecular and genetic mechanisms underlying
response and resistance associated with these different
groups. Unfortunately, cancer patients with similar
degrees of disease, who are equivalent in other important clinical respects, often differ widely with respect
to natural history and treatment outcome. The use of
clinical and morphological parameters to predict tumor
response or failure following local therapy are often
unsuccessful in large group studies and can not be used
reliably for treatment decisions for individual patients. The histopathological description of features
unique to neoplasia; enlarged nuclei, increased nuclear-
cytoplasmic ratios, hyperchromatism and mitotic index,
were the earliest indirect measures of cellular DNA
content. Since the 1950's, more quantitative and
objective measurements of DNA content have been determined by analytical cytometry techniques consisting at
present of sophisticated flow cytometric and image
analysis systems. These parameters have now been demonstrated to be important as prognostic indicators for
both pediatric and adult solid tumors. The parameter
most commonly determined is the DNA Index (DI) which
compares the cellular DNA content of a sample to a known
diploid standard which is arbitrarily defined as having
a DI = 1.0. The second parameter derived from a DNA
histogram is the percentage of cells synthesizing DNA,
or the %S Phase Fraction (SPF) which is an estimate of
the cells within the S-phase region of the DNA histogram.

PEDIATRIC SOLID TUMORS

Background

DNA content parameters have been shown to be useful
in the clinical evaluation and therapy of children with
solid tumors. The relationships of DNA content to prognosis and therapeutic outcome in pediatric malignancies,
particularily in terms of cytotoxic response, serve as a
model for defining similar relationships in the more
common adult solid tumors. Our discussion of the application of cellular DNA content parameters to the understanding of the clinical outcome of childhood malignancies will be restricted to three common solid tumors:
Neuroblastoma, rhabdomyosarcoma and osteosarcoma.

Neuroblastoma

Neuroblastoma is the most common extracranial solid
tumor seen in children. In this disease, DNA ploidy has
been shown to be associated with age, stage, histology,
response to chemotherapy, disease-free survival, and
overall survival (1-5). One of the earliest reports of
the prognostic and therapeutic significance of flow
cytometrically determined DNA content was in a report
from St. Jude's Research Hospital in 1984 (1). In 35
children less than 1 year of age, DNA ploidy of neuroblastoma correlated with stage of disease and response
to chemotherapy. This study demonstrated that DNA
diploid tumors were proportionately over-represented in
advanced stages of neuroblastoma. Additionally, in 23
patients treated with cyclophosphamide/doxorubicin
chemotherapy, 17/17 patients with DNA aneuploid tumors
responded, with complete responses observed in 15, while
0 of 6 patients with DNA diploid tumors responded despite similar treatment.

The relationship of DNA content and amplification
of the oncogene N-myc in neuroblastoma has more recently
been defined. N-myc amplification is uncommon in DNA

aneuploid neuroblastoma, and occurs in about half of the
DNA diploid, near diploid, and hypotetraploid tumors (5-
7). In a study involving 59 patients with neuroblastoma, DNA ploidy and N-myc expression were independent and
statistically significant prognostic variables and, in
multiparameter analysis, exceeded age and stage in their
predictive capabilities (6).

There are several apparent differences in the
relationship of the DNA content of neuroblastomas as
compared to adult solid tumors. In contrast to adult
malignancies where DNA aneuploid tumors have higher SPF,
DNA diploid neuroblastomas tend to have higher SPF (2).
The usual correlation of DNA ploidy with histology is
also reversed in neuroblastomas, with undifferentiated
neuroblastomas tending to be DNA diploid (2,3,5). In
adult solid tumors, DNA aneuploidy was initially associated with poor prognosis groups only, whereas, DNA
aneuploidy is associated with a better prognosis in
neuroblastomas (1-7). More recently, as will be discussed in the next section, the correlation of cytotoxic
response and DNA aneuploidy has been similarly confirmed
in a large number of adult tumors.

When combining DNA flow cytometry with an analysis
of N-myc expression and cytogenetic analysis, neuroblastomas separate into three distinct profiles that appear
to correlate with clinical presentation and outcome (8).
The first grouping of neuroblastomas occurs in younger
children, mainly infants less than one year. This group
is characterized by an increase in earlier stages (I,
II, IVS), and a non-random distribution of DI values
between 1.2 and 1.8, resulting in a near triploid DNA
content by flow cytometry. Cytogenetically, these
tumors lack chromosome 1p abnormalities, double minutes
and homogeneously staining regions (HSR). N-myc amplification is not seen. This group has a high response
rate to chemotherapy, and a very good prognosis for

disease-free and overall survival. The second group is
characterized by older age at diagnosis and more advanced clinical stages. By flow cytometry, DNA content
is diploid, near diploid, or hypotetraploid. Cytogenetic and molecular analysis, however, does not demonstrate
chromosome 1p abnormalities and N-myc amplification.
These patients often respond to chemotherapy, but relapse is common and most die of disease. The third
group is similar to the second clinically, having a near
diploid or hypotetraploid DNA content. Cytogenetic and
molecular analysis, however, reveal a high percentage of
chromosome 1p abnormalities, double minutes and HSR, and
N-myc amplification. This group does not respond to
chemotherapy and has a rapidly progressive clinical
course with a usually fatal outcome.

DNA content evaluations of neuroblastoma have
become an established guide to therapy (8). The greatest utility of DNA content analysis in neuroblastoma is
defining those with DNA aneuploidy, the better prognosis
group. In this group, shorter and less toxic therapy
can be administered with an expected high likelihood of
disease control. Patients with near diploid or tetraploid tumors can be further studied with molecular
techniques to detect N-myc amplification. Patients with
DNA aneuploid tumors can be expected to have >90% disease-free survival; tumors that are DNA diploid, near
diploid, or hypotetraploid without N-myc amplification
have an approximate 50% disease-free survival; and,
tumors without aneuploidy and with N-myc amplification,
a 10% disease-free survival (8).

Rhabdomyosarcoma

The descriptive information on DNA content parameter for rhabdomyosarcomas is limited. Two retrospective
studies using archival material for analysis, suggested
that nearly all rhabdomyosarcomas are DNA aneuploid

(9,10). A larger series from St. Jude's Hospital,
however, reported approximately 30% of rhabdomyosarcomas
as DNA diploid, and suggested a non-random distribution
of DNA indices based on histology as is seen in neuroblastoma (11). This series characterized the DNA content of 37 stage III and IV rhabdomyosarcomas. The
clinical characteristics of these patients were typical
for this disease. Twenty-four of the tumors were histologically classified as embryonal and 13 as alveolar.
DNA content analysis of all tumors revealed that about
one-third of patients had DNA diploid lesions regardless
of histology. In the group with DNA aneuploidy, however, marked differences in the distribution of DNA indices were seen based on histological classification. The
embryonal histology was associated with DI values between 1.1 and 1.8 in 15 of 16 DNA aneuploid tumors. In
contrast, of the 10 DNA aneuploid alveolar rhabdomyosarcomas, all had a near tetraploid DNA index. These data
suggest that in these histologic subtypes of rhabdomyosarcoma, different mechanisms are responsible for the
development of abnormal ploidy. In addition to this
association with histologic subtype, the presence of a
DNA aneuploid population appeared to correlate with both
response to chemotherapy and a favorable clinical outcome (11). In 20 patients with stage III disease, those
with DNA aneuploid populations in the hyperdiploid range
(DI = 1.1-1.8) had a higher response to chemotherapy,
and improved survival. This relationship between ploidy
and survival was statistically significant in both
univariate and multivariate analysis. In the alveolar
histologic subtype, the presence of tetraploidy ap-
proached significance for a favorable effect for overall
survival.

Osteosarcoma

The descriptive data of DNA ploidy in osteosarcoma

is more extensive than rhabdomyosarcoma, but is contradictory regarding the significance of DNA ploidy (12-
14). The largest experience reports that nearly all
high grade osteosarcomas are DNA aneuploid (12). No
correlation between histopathologic type of osteosarcoma
or tumor grade and DI was observed, nor could prognostic
significance for DNA aneuploidy or DI be demonstrated.
These data, in a smaller subset of patients, suggested
that SPF may have prognostic significance, with patients
having tumors with lower proliferative indices having a
better prognosis. The patients in this series had been
uniformly treated with surgery and interferon, but had
not received adjuvant chemotherapy.

A group of investigators from Boston has also
contributed considerable data regarding DNA content in
osteosarcomas. In 63 high grade osteosarcomas, Gebhardt
et al reported 46 to have DNA aneuploid populations
(13). When six patients who presented with metastatic
disease are excluded, DNA aneuploidy correlated with
improved disease-free survival (56% vs 18%, p<0.02) and
overall survival (69% vs 38%, p = 0.01). The patients
in this series were all treated with postoperative
adjuvant chemotherapy.

Finally, Look et al, examined the DNA content
parameters of 26 patients registered on the Multi Insti-
tutional Osteosarcoma Study (14). All patients in this
series were treated with the same postoperative adjuvant
chemotherapy. DNA content patterns were strongly correlated with both disease-free and overall survival in
this cohort. In 16 patients with tumors exhibiting a
near DNA diploid population of cells, two patients
relapsed for a disease-free survival rate of 88% at
three years. In contrast, 7 of 10 patients without a
near diploid population relapsed. DNA content pattern
of the osteosarcoma remained statistically significant
after adjustment for age, the only other significant

variable following univariate analysis.

Some of the difficulty in reconciling these studies results from different assumptions for DNA histogram interpretation. The Swedish group describes a tumor as DNA aneuploid when a population of cells with a DI <0.9 or >1.1 and a related G2M population is identified (12). While acknowledging that the DNA content of all osteo-sarcomas is abnormal, the Boston group characterizes osteosarcomas as aneuploid only when a population of cells outside the diploid or tetraploid range is identi-fied (13). The report of Look et al is unique in de-scribing a substantial portion of tumors without a near DNA diploid population (14). This finding may result from the tumor preparative methodology, or from examina-tion of cytospin preparations and interpretation re-garding the ratio of tumor to non-tumor cells. Adjuvant chemotherapy is another factor likely to influence the variability regarding the significance of DNA content in osteosarcoma. The full potential utility of DNA content analysis in osteosarcoma requires further study.

More recently, preoperative chemotherapy has become common in osteosarcoma. Several groups are interested in determining the relationship between changes in DNA ploidy as a result of preoperative chemotherapy and prognosis. Bosing examined the DNA content of 20 osteo-sarcomas treated with preoperative chemotherapy on the COSS 80/82 study and determined that fewer of the samples were DNA aneuploid, fewer had multiple popula-tions, and that the proliferative fractions were de-creased as compared to control samples in which preoper-ative chemotherapy was not given (15). Additionally, these investigators also suggested a relationship be-tween histopathologic necrosis at resection and the percentage of tumor that was DNA aneuploid.

The DNA content of 19 osteosarcomas was determined both at biopsy and following preoperative chemotherapy at resection the Boston group (16). In their series, all 8 patients demonstrating a good histologic response to preoperative chemotherapy had DNA aneuploid tumors at biopsy, and all lost the aneuploid population at surgical resection. Of 9 patients with poor response to preoperative therapy, only 2 had regression of DNA ploidy abnormalities.

Our experience at Wayne State University is similar in that an apparent increase in DNA diploidy in osteosarcoma following preoperative chemotherapy has been observed. At resection, 9 of 17 patients with high grade, localized osteosarcomas, has DNA diploid tumors. A relationship of DNA ploidy at surgery related to clinical outcome is suggested. While follow-up is short, 3 of the 4 relapses which have occurred are in patients in which DNA aneuploidy was found at resection. These findings suggest that the relationship of DNA ploidy following chemotherapy, as it relates to both disease- free interval and overall survival, requires further evaluation.

ADULT SOLID TUMORS
DNA Content Parameters

Aneuploid DNA content and/or high SPF have been predictive of poor survival associated with local and systemic relapse following surgical treatment in nearly every type of early staged, adult solid tumor investigated. These parameters have been well studied particularly for the common adult solid tumors of the breast (17), bladder (18), prostate (19), kidney (20), colon (21), and lung (22). On the other hand, more recent data would indicate that the DNA aneuploid components of advanced human tumors are those that are vulnerable to cytotoxic therapy. These include non-Hodgkin's lymphomas (23), advanced breast cancer (24), medulloblastomas (25), neuroblastomas (26), colon cancer treated with

radiotherapy (27), sarcomas treated with adjuvant chemotherapy (28), squamous cell lung cancer (29), and tran-
sitional cell cancer of the bladder (30). It is not
possible within the scope of this chapter to explore the
current status of the application of DNA content parameters in all human adult solid tumors, nor is it necessary since it is becoming clear that correlations established for one tumor are to a large extent applicable to
others. We will therefore confine the following discussion to human squamous cell carcinomas of the head and
neck (SCCHN), a tumor with which we have had considerable experience.

SCCHN

Most patients with SCCHN present with disease that
is incurable or is likely to fail definitive surgery
and/or radiotherapy (conventional therapy) (31). Furthermore, unacceptable functional and cosmetic deficits
occur in many patients that are cured with conventional
therapy. Progress in treating this disease depends on
the development of treatment strategies that identify
effective cytotoxic regimens for incurable patients
(induction trials), preserve organ function in resectable patients with advanced disease (organ preservation
trials) and reduce local or distant recurrences following conventional therapy (adjuvant trials) (32,33). The
success of these strategies will require the identification of patient groups that will benefit from cytotoxic
therapy.

Survival advantages have been demonstrated in
induction trials when clinical complete responses (CR)
have resulted in the eradication of microscopic disease
(34). Nevertheless, the majority of CR do not attain
this status and an additional 50% develop gross resistant disease (partial and non-responders) during treatment (31-33). Sequential response patterns for patients

with advanced SCCHN treated with multi-modality cytotoxic therapy have also been shown to be an important
consideration in designing clinical trials and suggest
that the choice of cytotoxic treatment modality and its
timing may impact on overall tumor response (35).
Knowledge of which tumors or components are responsive
to each of these modalities would be extremely useful in
selecting initial cytotoxic therapy and designing sequential therapies in multi-modaility approaches to
management. The reliable pretherapeutic prediction of
cytotoxic response would permit the selection of patients with advanced resectable disease who are likely
to benefit from organ preservation trials and would
otherwise lose important organ function as a result of
conventional therapy.

Patients with advanced and even early stage, resectable SCCHN continue to fail conventional therapy at
an unacceptable rate (36). Several phase III clinical
trials have now suggested that the addition of cytotoxic
therapy in the neoadjuvant, adjuvant or maintenance
setting may reduce distant metastases as a form of tumor
relapse (37-39). Unfortunately, the patients most
likely to benefit from the addition of cytotoxic therapy
can not be reliably identified. This diminishes the
ability to detect the impact of adjuvant therapy if
present, increases the number of patients required for
such a study and subjects patients with a low risk of
relapse to the toxicities of adjuvant cytotoxic therapy.

Cellular DNA Content Parameters in SCCHN

Most studies of SCCHN, with one exception (40),
report survival advantages associated with DNA diploid
tumors for patients with early stage, resectable SCCHN
(41-44). These have all been retrospective studies. We
have prospectively correlated cellular DNA parameters
using fresh specimens and whole-cell flow cytometry in

200 patients with <u>advanced</u> resectable SCCHN treated with surgery as initial therapy. With a <u>minimum</u> follow-up of two years for all patients, local-regional recurrence rates are higher, and both disease-free and absolute survival are lower for DNA aneuploid tumors (45-47). At two years, 80% of the patients with DNA diploid tumors, versus 49% for DNA aneuploid tumors, were either alive without disease or had died disease-free.

We have also demonstrated that tumor growth characteristics and stromal-inflammatory response patterns are significantly different for DNA aneuploid and DNA diploid tumors in patients with advanced SCCHN (48). These patterns indicate that negative surgical margins are more likely to be achieved by the surgeon and detected by the pathologist in patients with DNA diploid tumors. This suggests an explanation for the high local relapse rates for DNA aneuploid tumors following surgery as initial therapy. These findings, and the fact that DNA diploid tumors respond poorly to cisplatinum combination chemotherapy as discussed in the following section, make advanced resectable patients with DNA diploid tumors poor choices for adjuvant clinical trials employing such regimens. DNA content parameters, therefore, have potential for selecting or stratifying patients with advanced, resectable tumors that are entered onto clinical trials employing various adjuvant cytotoxic regimens.

Previous to our work, no published studies had related cellular DNA content parameters to chemotherapy response in SCCHN tumors; a few retrospective studies of survival for patients with early stage tumors following radiation have been reported (49,50). Similar CR rates have been reported for DNA diploid and aneuploid SCCHN tumors (73% <u>vs</u> 65%, respectively) following radiotherapy although survival was slightly better for patients with DNA diploid tumors (50% <u>vs</u> 38%) (49). Conversely, Franzen <u>et al</u>, reported survival advantages for patients with low stage DNA aneuploid tumors following radiotherapy (50). Serial measurements of both DNA ploidy and cytokinetics during radiotherapy demonstrated that radioresistant tumors become more well differentiated, with lower percent labeled mitoses and less DNA aneuploid DNA content (51). Cell lines established from metastatic or recurrent SCCHN tumors have a lower DI and chromosome counts than those from untreated specimens (52). Several recent pilot studies of the correlation of response to radiotherapy and DNA content parameters in patients with SCCHN have also been contradictory (53).

We initially reported that the DNA aneuploidy, mean DI and associated SPF were significantly lower in specimens from patients presenting with recurrent tumor than those of previously untreated patients (54). Since most of these recurrences followed treatment with radiotherapy and/or chemotherapy, these data provided preliminary evidence for the cytotoxic vulnerability of DNA aneuploid subpopulations. Prospective, serial determinations of DNA content parameters in 200 advanced unresectable patients undergoing cisplatin-containing combination cytotoxic therapy indicate that at both the clinical and microscopic level, resistant tumors or tumor subpopulations are nearly always DNA diploid, whereas DNA aneuploid tumors or DNA aneuploid subpopulations of SCCHN are those portions that are responsive (46,47,55). The correlation of DNA aneuploid SCCHN tumors and cytotoxic response has recently been substantiated by two other independent reports (56,57).

Our data indicate that in approximately half of the patients presenting with tumors containing both DNA aneuploid and DNA diploid components, tumor is absent at the microscopic level following cytotoxic therapy. In these instances, the DNA diploid component of the pretreatment specimen is unlikely to be tumor considering the data on response of pure DNA diploid tumors. This suggests the possibility that microscopic eradication of disease is attained only when pure DNA aneuploid tumors are present. In vitro evidence indicates that pure DNA diploid and mixed DNA diploid-aneuploid cultures do transform to pure, stable DNA aneuploid lines (58). We have developed and validated a single cell immunofluorescent flow cytometric assay for tumor specific cytokeratin content which are not present in non-tumor DNA diploid elements (CAM 5.2, BD 7650). A prospective, multiparameter analysis of 50 tumors indicated that 92% of histologically confirmed, pure DNA diploid tumors, are cytokeratin positive (50). DNA aneuploid components of mixed DNA ploidy specimens reacted with this monoclonal 94% of the time whereas only 62% of the corresponding DNA diploid components were reactive. This confirms previous indirect evidence that pure DNA aneuploid SCCHN tumors exist. This multiparameter technique documents the existence of pure DNA aneuploid tumors, and will permit the pretherapeutic identification and selection of such patients for clinical trials employing cytotoxic therapy. The flow cytometric screening of tumor cellular DNA content parameters has potential clinical application for monitoring SCCHN patients with advanced disease during induction and adjuvant trials as well as aiding in the selection of candidates for organ preservation trials.

CONCLUSIONS

When properly optimized and validated methodology is developed, DNA content parameters can be extremely useful in predicting the natural history and treatment outcome for patients with pediatric and adult solid tumors. The clinical and therapeutic differences associated with DNA content parameter groups reflect significant differences in underlying biological mechanisms at the molecular and genetic level. The existence of differences in cytogenetics between DNA diploid and aneuploid tumors is self-evident, but, due to the complexity of solid tumor cytogenetics, have been difficult to describe or study. At the molecular genetic level, activation of certain oncogenes, deactivation or mutation of tumor suppressor genes, and alterations in tumor growth factor systems are currently being correlated with differences in tumor DNA content. DNA content parameters therefore serve as an important intermediate marker for the cytogenetic and molecular changes in human solid tumors that ultimately determine their natural history and response to therapy.

REFERENCES

1. Look AT, Hayes FA, Nitschke R, et al: Cellular DNA content as a predictor of response to chemotherapy in infants with unresectable neuroblastoma. N. Engl. J. Med. 311:231-235, 1984.
2. Gansler T, Chatten J, Varello M, et al: Flow cytometric DNA analysis of neuroblastoma. Correlation with histology and clinical outcome. Cancer 58: 2453-2458, 1986.
3. Oppedal BR, Storm-Mathisen I, Lie SO, Brandtzaeg P: Prognostic factors in neuroblastoma. Clinical, histopathologic, and immunohistochemical features and DNA ploidy in relation to prognosis. Cancer 62:772-780, 1988.
4. Naito M, Iwafuchi M, Ohsawa Y, et al: Flow cytometric DNA analysis of neuroblastoma: Prognostic significance of DNA ploidy in unfavorable group. J. Ped. Sur. 26:834-837, 1991.
5. Cohn SL, Rademaker AW, Salwen HR, et al: Analysis of DNA ploidy and proliferative activity in relation to histology and N-myc amplification in neuroblastoma. Am. J. Path. 136:1043-1051, 1990.
6. Bourhis J, DeVathaire F, Wilson GD, et al: Combined analysis of DNA ploidy index and N-myc genomic content in neuroblastoma. Cancer Res. 51:33-36, 1991.
7. Look AT, Hayes FA, Shuster JJ, et al: Clinical relevance of tumor cell ploidy and N-myc gene amplification in childhood neuroblastoma: A pediatric oncology group study. J. Clin. Oncol. 9:581-591, 1991.
8. Brodeur GM: Neuroblastoma: Clinical significance

of genetic abnormalities. Cancer Surveys 9:673-685, 1990.

9. Boyle ET Jr, Reiman HM, Kramer SA, et al: Embryonal rhabdomyosarcoma of bladder and prostate: Nuclear DNA patterns studied by flow cytometry. J. Urol. 140:1119-1121, 1988.

10. Molenaar WM, Dam-Meiring A, Kamps WA, Cornelisse CJ: DNA-Aneuploidy in rhabdomyosarcomas as compared with other sarcomas of childhood and adolescence. Hum. Pathol. 19:573-579, 1988.

11. Shapiro DN, Parham DM, Douglass EC, et al: Relationship of tumor-cell ploidy to histologic subtype and treatment outcome in children and adolescents with unresectable rhabdomyosarcoma. J. Clin. Onc. 9:159-166, 1991.

12. Bauer HCF: DNA Cytometry of Osteosarcoma. Acta Orthop. Scand. 59 (suppl. 228):1-39, 1988.

13. Gebhardt MC, Lew RA, Bell RS, et al: DNA ploidy as a prognostic indicator in human osteosarcoma. Proc. Eur. Musculoskeletal Onc. Soc. Bologna, Italy, suppl. 1:18-21, 1990.

14. Look AT, Douglass EC, Meyer WH: Clinical importance of near-diploid tumor stem lines in patients with osteosarcoma of an extremity. N. Engl. J. Med. 318:1567-1572, 1988.

15. Bosing T, Roessner A, Hiddemann W, et al: Cytostatic effects in osterosarcomas as detected by flow cytometric DNA analysis after preoperative chemotherapy according to the COSS 80/82 protocol. J. Cancer Res. Clin. Oncol. 113:369-375, 1987.

16. Baldini N, Gebhardt MC, Springfield DS, et al: Effect of preoperative chemotherapy on nuclear DNA content in osteosarcoma. Proc. of Eur. Musculoskeletal Onc. Soc., Bologna, Italy, suppl. 1:22-24, 1990.

17. McDivitt RW, Stone KR, Craig B, et al: A proposed classification of breast cancer based on kinetic information. Cancer 57:269-276, 1986.

18. Frankfort OS, Huben RP: Clinical applications of DNA flow cytometry for bladder tumors. Urol. Supp. 23:29-34, 1984.

19. Zetterberg A, Esposti PL: Prognostic significance of nuclear DNA levels in prostatic carcinoma. Scand. J. Urol. Nephrol. 55:53-58, 1980.

20. Otto LL, Baisch H, Hulano H, Kloppel G: Tumor cell deoxyribonucleic acid content and prognosis in human renal carcinoma. J. Urol. 132:237-239, 1984.

21. Wolley RC, Schreiber K, Koss LG, et al: DNA distribution in human colon carcinoma and its relationship to clinical behavior. J. Natl. Cancer Inst. 69:15-22, 1982.

22. Volm M, Hahn EW, Mattern J, Vogt-Moykopf I: Independent flow cytometric prognostic factors for the survival of patients with non-small cell lung can-

cer: A five year follow up study. Proc. AACR 29:26, 1988.

23. Woodbridge TN, Grierson HL, Pierson JL, et al: DNA aneuploidy and low proliferative activity predict a favorable clinical outcome in diffuse large cell lymphoma. Proc. AACR 28:131, 1987.

24. Brifford M, Spyratos F, Tubiana-Hulin M, et al: Analysis of breast carcinoma response to chemotherapy by sequential cytopunctures: Influence of DNA analysis, morphological changes and histology on tumor regression. Proc. ASCO 7:34, 1988.

25. Tomita T, Yasue M, Englehard HH, et al: Flow cytometric DNA analysis of medulloblastomas. Prognostic implications of ploidy. Cancer 61:744-749, 1988.

26. Oppedal BR, Strom-Mathisen I, Lie SO, Brandtzaeg P: Prognostic factors in neuroblastomas: Clinical, histopathologic, and immunohistochemical featues and DNA ploidy in relation to prognosis. Cancer 62:772-780, 1988.

27. Meyer JS, Priolea PG: S-phase fractions of colorectal carcinomas related to pathologic and clinical features. Cancer 48:1221-1228, 1981.

28. Bell RS, Mankin HJ, Gebhardt MC, Lee R: Disease free survival and tumor ploidy in high grade osteosarcoma. Proc. 10th Annual Meeting Cell Kinetics Society, Santa Fe, New Mexico, p. 30, 1986.

29. Rosell R, Gomez-Codina J, Camps C, et al: Favorable outcome and aneuploidy reversion following neoadjuvant chemotherapy in stage IIIA non-small cell lung cancer. Proc. ASCO 11:954, 1992.

30. Jacobsen AB, Berner A, Juul M, et al: Prognostic significance of deoxyribonucleic acid flow cytometry in muscle invasive bladder carcinoma treated with preoperative radiation and cystectomy. J. Urol. 147:34-37, 1992.

31. Al-Sarraf M , Kish J, Ensley JF: Head and neck cancer. The Wayne State University Experience with Adjuvant Chemotherapy. Hem. Onc. Clinics North Amer. 5:687-700, 1991.

32. Ensley JF, Maciorowski Z, Kish JA, Al-Sarraf M: The significance of pretreatment identification of prognostically important subgroups of squamous cell cancer of the head and neck. In: Scientific and Clinical Perspective of Head and Neck Cancer Management, Strategies for Cure, J Jacobs, M Al-Sarraf, J Crissman, F Valeriote (ed), Elsevier Inc., NY, pp. 35-56, 1987.

33. Ensley J, Kish J, Tapazoglou E, et al: The justification and strategies for the continued intensification of induction regimens in patients with advanced, untreated head and neck cancer. In: Head and Neck Oncology Research. Proceedings of the Second International Research Conference on Head

and Neck Cancer, 1987, GT Wolf, T Carey (ed), Kugler Pub., Amsterdam, pp. 313-321, 1988.

34. Al-Kourainy K, Kish JA, Ensley JF, et al: Achievement of superior survival for histologically negative vs histologically positive clinically complete responders to cis-platinum combinations in patients with locally advanced head and neck cancer. Cancer 59:233-238, 1987.

35. Ensley JF, Kish JA, Jacobs J, et al: Incremental improvements in median survival associated with degree of response to adjuvant chemotherapy in patients with advanced squamous cell cancer of the head and neck. In: Adjuvant Therapy of Cancer, Vol. IV, SE Jones, SE Salmon (ed), Grune & Stratton, Orlando, pp. 117-126, 1984.

36. Looser KG, Shah JP, Strong EW: The significance of "positive" margins in surgically resected epidermoid carcinomas. Head and Neck Surgery 1:107-111, 1978.

37. Ervin TJ, Clark JR, Weichselbaum RR, et al: An analysis of induction and adjuvant chemotherapy in multidisciplinary treatment of squamous-cell carcinoma of the head and neck. J. Clin. Oncol. 5:10-20, 1987.

38. Wolf GT, et al: Induction chemotherapy plus radiation compared with surgery plus radiation in patients with advacned laryngeal cancer. N. Engl. J. Med. 324:1685-1690, 1991.

39. Adjuvant chemotherapy for advanced head and neck squamous carcinoma: Final report of the Head and Neck Contracts Program. Cancer 59:301-311, 1987.

40. Goldsmith MM, Cresson DH, Postma DS, et al: The significance of ploidy in laryngeal cancer. Am. J. Surg. 152:396-402, 1986.

41. Sickle-Santanello BJ, Farrar WB, Dobson JL, et al: Flow cytometry as a prognostic indicator in squamous cell carcinoma of the tongue. Am. J. Surg. 152:393-395, 1986.

42. Kokal RL, Gardine K, Sheibani IW, et al: Tumor DNA content as a prognostic indicator in squamous cell carcinoma of the head and neck region. Proc. ASCO 7:149, 1988.

43. Gussack GS, Donelly K, Hester R, Dowling E: Flow cytometric DNA analysis of laryngeal carcinomas. In: Head and Neck Oncology Research. Proceedings of the Second International Research Conference on Head and Neck Cancer, 1987, GT Wolf, T Carey (ed), Kugler Pub., Amsterdam, pp. 241-249, 1988.

44. Kaplan AS, Caldarelli DD, Chacho MS, et al: Retrospective DNA analysis of head and neck squamous cell carcinoma. Arch. Otolaryngol. Head and Neck Surg. 112:1159-1162, 1986.

45. de Braud F, Ensley JF, Hassan M, et al: Prospective correlation of clinical outcome in patients

with advanced, resectable squamous cell carcinomas of the head and neck (SCCHN) with DNA ploidy from fresh specimens. Proc AACR 30:1046, 1989.

46. Ensley J, Maciorowski Z, Pietraskiewicz H, et al: Prognostic impact of cellular DNA parameters in neoadjuvant and adjuvant trials of patients with head and neck (H&N) cancer. In: Adjuvant Therapy of Cancer, Vol. VI, SE Salmon (ed), Grune & Stratton, Orlando, pp. 101-108, 1990.

47. Ensley JF, Maciorowski Z, Pietraszkiewicz H, et al: Methodology and clinical applications of cellular DNA content parameters determined by flow cytometry in squamous cell cancers of the head and neck. Cancer Treat. Rep. 52:225-242, 1990.

48. Sakr W, Hassan M, Hassan M, et al: DNA quantitation and histologic characteristics of squamous cell carcinomas of the upper aerodigestive tract. Arch. Pathol. Lab. Med. 113:1009-1014, 1989.

49. Holm LE: Cellular DNA amounts of squamous cell carcinomas of the head and neck region in relation to prognosis. Laryngoscope 92:1064-1069, 1982.

50. Franzen G, Olofsson J, Tytor M, et al: Preoperative irradiation in oral cavity carcinoma. A study with special reference to DNA pattern, histological response and prognosis. Acta Oncol. 26:349-355, 1987.

51. Nervi C, Badaracco G, Morelli M, Starace G: Cytokinetic evaluation in human head and neck cancer by autoradiography and DNA cytofluorometry. Cancer 45:452-459, 1980.

52. Roa RA, Carey TE, Passamani PP, et al: DNA content of human squamous cell carcinoma lines. Analysis by flow cytometry and chromosome enumeration. Arch. Otolaryngol. 111:565-575, 1985.

53. Proceedings of the Third International Head and Neck Oncology Research Conference, Las Vegas, 1990.

54. Ensley JF, Maciorowski Z, Hassan M, et al: Cellular DNA parameters in untreated and recurrent squamous cell cancers of the head and neck. Cytometry 10:334-338, 1989.

55. Ensley J, Maciorowski Z, Hassan M, et al: Prospective correlations of flow cytometry (FCM) DNA parameters (DNA Index and % SPF) and cytotoxic response in previously untreated patients with advanced squamous cell cancers of the head and neck. Proc. AACR 29:105, 1988.

56. Tennvall J, Wennergerg J, Anderson H, et al: DNA analysis as a predictor of the outcome of induction chemotherapy in advanced head and neck carcinomas. Arch. Otolaryngol. - Head and Neck Surgery, In press.

57. Cooke LD, Cooke TG, Bootz F, et al: Ploidy as a prognostic indicator in end stage squamous cell carcinoma of the head and neck region treated with

cisplatinum. Br. J. Cancer 61:759-762, 1991.

58. Ensley J, Maciorowski Z, Pietraszkiewicz H, Reed M: DNA ploidy transformations in new head and neck cancer cell lines (SCCHN): Implications for cyto- toxic therapy. Proc. AACR 31:23, 1990.

59. Ensley J, Maciorowski Z, Pear A, et al: The selec- tion of patients for clinical trials with squamous cell cancers of the head and neck (SCCHN) based on the existence of pure DNA aneuploid tumors docu- mented by multiparameter flow cytometry. Proc. ASCO, 1993, In press.

12

DIFFERENTIATION OF HUMAN B-CELL TUMORS: A PRECLINICAL
MODEL FOR DIFFERENTIATION THERAPY

Ayad M. Al-Katib and Ramzi Mohammad

INTRODUCTION

B cell tumors in man include a group of heterogenous diseases with varying natural histories and responsiveness to therapy. Classic examples of B cell tumors
are the chronic lymphocytic leukemia (CLL), Burkitt's
lymphoma and multiple myeloma. These tumors express the
conventional B cell marker, that is, surface and/or
cytoplasmic immunoglobulins. Malignant transformation,
however, can affect precursors of the "mature" B lymphocytes as exemplified by the non-T cell acute lymphoblastic leukemia (ALL). Such cases demonstrate immunoglobulin gene rearrangements and react with monoclonal antibodies to B cell differentiation antigens. B cell
tumors, therefore, represent a spectrum of disorders
extending from the immature "stem cell" to the most
mature "plasma cell" of the B lineage. It has been long
hypothesized that disturbance in the differentiation
pathway is important in the pathophysiology of malignancy (1). Phenotypic analysis of the B cell lineage has
identified a malignant counterpart phenotype for each
stage of the normal B cell differentiation pathway (3).
Each tumor then, represents a monoclonal population of
cells that are "frozen" or "arrested" at a certain stage
of maturation.

Differentiation therapy is based on the principle
that administration of chemical or biological agents can

induce the tumor cells to overcome the maturation block.
In this way, the imbalance between the proliferation and
differentiation pathways is restored. Recently, Sachs
has postulated that differentiating agents can restore
such a balance by bypassing the genetic defects that led
to the malignant transformation (2). We feel that B
cell tumors provide a suitable model for the development
of differentiation therapy in cancer for many reasons.
These tumors represent discrete stages of B cell differentiation. Often the cells are easily accessible from
blood, bone marrow or superficial lymph nodes and are
amenable to multiparameter analysis by flow cytometry
and other techniques. The availability of well-charact-
erized, established cell lines is an added resource to
facilitate research on these tumors. Over the past few
years, we have established a number of human B cell
tumor cell lines representing the clinical spectrum of
these disorders. This chapter summarizes the working
schema of B cell differentiation used in our laboratory
and the utility of our B cell lines as models for B cell
differentiation studies.

A WORKING SCHEMA FOR B CELL DIFFERENTIATION

Unlike the granulocytic series where each stage of
differentiation has characteristic light microscopic
features, morphology is not a reliable measure of the
differentiation state in the lymphoid lineage. A number
of monoclonal antibodies to B cell differentiation
antigens have been developed during the past two decades. Some of these antibodies are not only lineage-
associated but also stage-restricted. The gain or loss
of such markers in response to exogenous agents can
provide an objective measure of a change in the differentiation state of the lymphocytes. Based on this
assumption, a number of hypothetical models for B cell
differentiation have been published by different labora-

tories (3,4). Shown in Fig. 1 is the schema currently
used in our laboratory. The range of reactivity of each
marker is hypothetically determined based on expression
on fresh cases of B cell tumors, established cell lines
from different tumors and <u>in vitro</u> induction experiments
on normal and malignant B cell precursors (5-9). We
have shown that the HLA-DQ/C expression is more restricted than that of HLA-DR (10) and that its expression can be induced on pre-B cell lines by the phorbol

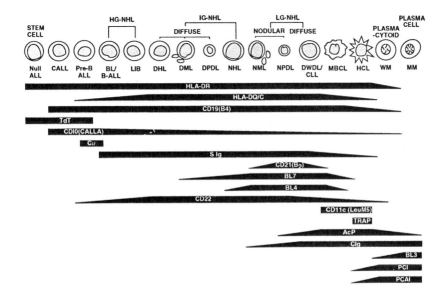

Figure 1. A hypothetical schema of B cell differentiation. Stages are arranged on the top from the most
immature (stem cell) on the far left to the most mature
(plasma cell) on the far right. ALL = Acute lymphoblastic leukemia; CALL = Common ALL; BL = Burkitt's lymphoma; LIB = Large cell immunoblastic lymphoma; DHL =
Diffuse histiocytic lymphoma; DML = Diffuse mixed lymphoma; DPDL = Diffuse poorly differentiated lymphocytic
lymphoma; NHL = Nodular histiocytic lymphoma; NML =
Nodular mixed lymphoma; NPDL = Nodular poorly differentiated lymphocytic lymphoma; DWDL/CLL = Diffuse well
differentiated lymphocytic lymphoma/Chronic lymphocytic
leukemia; MBCL = Monocytoid B cell lymphoma; HCL = Hairy
cell leukemia; WM = Waldenstrom's macroglobulinemia; MM
= Multiple myeloma; HG-NHL = High grade non-Hodgkin's
lymphoma; IG-NHL = Intermediate grade NHL; LG-NHL = Low
grade NHL.

ester 12-O-tetradecanoylphorbol-13-acetate (TPA) (11).
The phorbol esters are known to induce B cell differentiation as measured by the synthesis and secretion of
immunoglobulins (5,12-14). CD19, like HLA-DR, is considered a pan B marker being expressed on all stages of
B cell maturation except the plasma cells and the earliest stage (stem cells) (15). The expression of the
terminal deoxynucleotidyl transferase (TdT) in B cell
lineage is restricted to the surface immunoglobulin
(SIg)-negative ALL (16). We have previously reported
that treatment of the ALL Reh cell line with TPA results
in loss of the TdT expression (17). While the expression of CD10 is equated with ALL, and hence the name
common acute lymphoblastic leukemia antigen (CALLA), it
is also expressed in other stages of the B cell lineage,
particularly the follicular (nodular) lymphomas (3,18).
The expression of cytoplasmic IgM heavy chain $(c\mu)$ is
synonymous with the pre-B ALL (19). This is believed to
precede the synthesis and expression of the full Ig
molecule on the surface (SIg), which in turn precedes
the expression of the full Ig molecule in the cytoplasm
(CIg).

Certain surface antigens are expressed at intermediate stages of the B cell differentiation pathway.
Some of these are the CD21(B2) (20), BL4 (4) and BL7
(21). Such markers are very useful in assessing the
transition of B cells along the differentiation pathway.
During this process, the intensity of expression rises
to a peak and fades away gradually. Therefore, during
induction experiments, one may see an initial increase
or decrease in the expression of these markers based on
the initial state of differentiation of the cell. The
CD11c and CD22 are two markers that were first reported
in 1985 (22). While CD11c is expressed on monocytes and
CD22 on B lymphocytes, the co-expression of the two
markers was initially thought to be specific for hairy

cell leukemia (HCL). Since then, however, a new subset of non-Hodgkin's lymphoma, the monocytoid B cell lymphoma (MBCL), has been described that also co-expresses these two markers (23,24). Both HCL and MBCL also express acid phosphatase (AcP). However, such expression can be inhibited by tartrate (tartrate sensitive) in MBCL but not in HCL (tartrate resistant, TRAP) (25). Because of its usefulness, we have added the AcP and TRAP to our schema.

The conventional markers of the most mature B cells, the plasma cells, are the intense CIg expression and secretion. In addition to that, certain monoclonal antibodies (BL3, PC1, PCA) are believed to react with surface antigens that are associated or restricted to the Ig-secreting stage (26,27). We have, therefore, included these reagents in our schema as markers of late stages of B cell maturation.

MALIGNANT B CELL LINES AS MODELS FOR B CELL DIFFERENTIATION STUDIES

The availability of fresh tumor specimens from patients and the number of cells obtained from such specimens are often the limiting factors for extensive differentiation studies. In cell types, like the B cells, where staining with several monoclonal antibodies is required for objective evaluation of the change in differentiation state, a large number of cells is needed. Additional numbers of cells is required if other parameters, like cell cycle analysis and oncogene expression, are included. Moreover, fresh tumors often contain normal reactive cells, red blood cells, fat and debris that have to be removed. Once the patient is treated, the same tumor may not be available for further experimentation. Sometimes, repeating certain experiments becomes crucial before conclusive evidence is drawn. These limitations can be circumvented by the

availability of established and well characterized tumor
cell lines. Such lines provide a continuous source of
usually homogeneous populations of cells for investigation. For cell lines to be representative of the original tumors from which they were established, the major
phenotypic and karyotypic features have to be preserved.
Some of the B cell lines reported in the literature have
been established through Epstein-Barr Virus (EBV) infection (28). Although such lines are probably of malignant origin, it remains unclear whether the incorporated
EBV genome has altered their biological characteristics.
Because of this concern, the EBV-transformed cell lines
are not ideal for differentiation studies.

Since 1986, we have successfully established seven
cell lines from different human B cell tumors. All of
them were established without the aid of exogenous
mitogens, growth factors or viral transformation and all
are EBV-negative. Not every attempt at establishing a
permanent cell line from human tumor was successful.
The seven cell lines established were out of approximately 70 attempts giving a success rate of 10%. There
are no clear predictive factors of successful establishment of a cell line in our hands. However, it seems
that tumor cells derived from serous effusions have a
better chance of continuous growth in vitro since five
of the cell lines were established from either a pleural
effusion or ascites. This is perhaps due to the fact
that such cells may have already been adapted to growth
in liquid medium resembling that of the in vitro culture
system. Once the fresh specimen is received in the
laboratory, the mononuclear cells are isolated by Ficoll-Hypaque (Pharmacia Fine Chemicals, piscataway, NJ)
density centrifugation and washed twice with either
phosphate buffered saline (PBS) or Hank's balanced salt
solution (HBSS).

These cells are then plated at varying densities

ranging from 0.5-10x10^6 cells/ml in RPMI-1640 medium
supplemented with 20% fetal bovine serum (Hyclone,
Logan, UT). In our experience, the cell density does
not play a role in the successful establishment of a
cell line. Moreover, we have also used varying concentrations of FBS ranging from 10 to 30% with no superior-
ity of any particular concentration.

Fig. 2 shows the various types of B cell tumors
together with the stages (top portion similar to that of
Fig. 1) with available representative cell line(s) for
the corresponding tumor type below. Cell lines starting
with WSU are those established in our laboratory at
Wayne State University. All of these cell lines, with
the exception of WSU-CLL have been published (29-34).
It should be stressed that the schema shown in Fig. 2 is
hypothetical. The nodular (follicular) lymphomas are
believed to be more "mature" than the diffuse lymphomas.
However, the difference between the small cell and large
cell lymphomas (or the poorly differentiated lymphocytic
and the histiocytic, in the Rappaport classification)
may be related to transformation rather than differentiation (35). The establishment of new B cell tumor lines
of stages not represented in Fig. 2 may help us under-
stand the differentiation-stage relationship between the
different diseases. As new knowledge is gained, the
schema may change accordingly. The established cell
lines together with schema of markers shown in Fig. 1
are proving to be useful tools in the understanding of
this complex subject. Through the use of a short-term
in vitro culture system, we and others have shown that
such B cell tumors are amenable to differentiation by
certain chemical or biological agents.

INDUCTION OF B CELL DIFFERENTIATION IN VITRO:
METHODOLOGY

The design of a typical induction experiment in

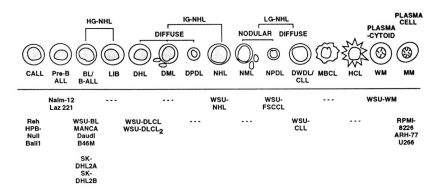

Figure 2:   Representative cell lines for B cell tumors
(bottom).   Stages of B cell tumors (on top) are same as
in Fig. 1.

vitro is shown in Fig. 3.   The source of malignant B
cells is either fresh specimens from patients or established cell lines.   Peripheral blood, bone marrow or
lymph nodes are the most common sources although we have
also used malignant effusions.   Following Ficoll-Hypaque
density centrifugation and washing, the cell suspension
is divided equally into a number of aliquots depending
on the number of treatments planned plus a control.
Cells in each aliquot are then distributed equally into
a number of T-25 flasks (Falcon Labware, Oxnard, CA)
depending on the number of time points in the experiment.   At each time point, one flask from the control
and each treatment group are removed, cells washed with
HBSS and evaluated for evidence of differentiation as
shown in Table 1.

PROTEIN KINASE C (PKC) ACTIVATORS INDUCE B CELL DIFFERENTIATION

Of the PKC activators, the phorbol esters are the
best known agents that induce B cell differentiation in
vitro.   Hokland et al have shown that normal human pre-B
cells (isolated from fetal bone marrow) can be induced
to differentiate in vitro by phorbol myristic acid (5).

Table 1
In Vitro Induction of B cell Differentiation:
Analysis of Cells at Each Time Point

1. Growth inhibition/ cytoxicity	- Cell count (Hemocytometer) - Viability (Trypan blue dye exclusion) - Cell cycle analysis (flow cytometry)
2. Morphological changes	- Cytocentrifuge smear for light microscopy - Electron microscopy
3. Phenotypic changes	- Indirect immunofluorescence (IF) staining with B cell monoclonal antibodies - IF staining for surface (SIg) and cytoplasmic (CIg) immunoglobulins
4. Functional assays	- Immunoglobulin Secreting Cells (ISC) - Immunoglobulin secretion in the culture supernatant - Latex bead phagocytosis (for HC and MBL stages*)
5. Enzymatic assays	- Immunohistochemical stains of cytocentrifuge smears for various enzymes, e.g. Terminal transferase (TdT), acid phosphatase with and without tartrate inhibition
6. Changes in gene expression	- PCR, northern blots or flow cytometric assays for oncogenes or their products (myc, fos, bcl2, etc.), or for mdrl gene
7. Changes in protein expression	- Two dimensional protein gel electrophoresis (2D-PAGE)

*HC = hairy cells; MBL = monocytoid B lymphocytes

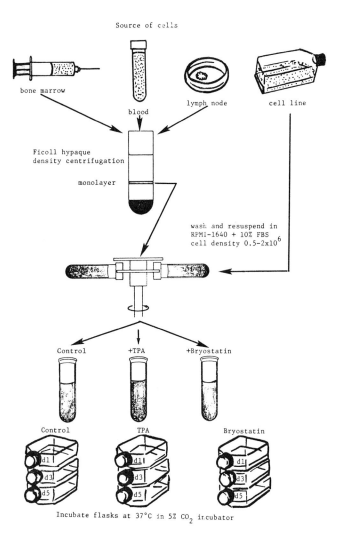

Source of cells

bone marrow

blood

lymph node

cell line

Ficoll hypaque
density centrifugation

monolayer

wash and resuspend in
RPMI-1640 + 10% FBS
cell density $0.5-2 \times 10^6$

Control   +TPA   +Bryostatin

Control   TPA   Bryostatin

d1   d3   d5

Incubate flasks at 37°C in 5% CO_2 incubator

Figure 3: Design of induction of B cell differentiation
expermiment in vitro.

Interestingly, several laboratories have shown that
phorbol esters can also induce in vitro differentiation
of a variety of human B cell tumors. Totterman et al
found that 12-O-tetradecanoylphorbol 13-acetate (TPA)
can induce CLL cells to an Ig-secreting plasmacytoid
stage (36). Cells from a patient with common ALL were
also found to differentiate in vitro in response to TPA,

acquiring a B cell associated antigen and decreasing expression of TdT, as well as the production of cyto-plasmic and ultimately surface Ig's (12). In our labo-ratory, we were able to demonstrate that TPA induces differentiation of B cell tumors to a specific stage along the B cell differentiation pathway. For example, the common ALL line, Reh, is induced to a monocytoid B lymphocyte (MBL) stage (17) while CLL cells are induced to a hairy cell (HC) stage (37). MBL and HC are closely related with common features including adherence to plastic surface and co-expression of CD22 and CD11c. Both cell types also express acid phosphatase. However, this expression is inhibited by tartrate in MBL (i.e., tartrate sensitive) but not in HC (tartrate resistant).

In addition to the changes in the phenotypic and enzymatic profiles during B cell differentiation, a number of other cellular parameters have been evaluated. For example, as in most systems, we have shown that TPA-induced B cell differentiation is usually accompanied by cessation or inhibition of growth (17). Such a determi-nation can be made by simple cell counting and viability using a hemocytometer or by flow cytometric analysis of the cell cycle. Flow cytometry offers an added advan-tage of concurrent evaluation of a second parameter in a dual staining procedure. Using such a technique, we have shown that TPA induced growth inhibition and down regulation of c-myc oncogene expression in HL-60 but not Daudi Burkitt's lymphoma cells (38).

It is assumed that exogenous agents, like TPA, induce B cell differentiation through a genetic change. As such, the phenotypic, functional, morphological and enzymatic changes can be viewed as end results or mani-festations of differentiation. In an attempt to under-stand the mechanism of B cell differentiation, we have analyzed the changes in total cellular proteins of two B cell non-Hodgkin's lymphoma lines using 2D-PAGE. Our study revealed that similar to the phenotypic markers, there seems to be an orderly expression of certain cellular proteins detected by 2D-PAGE (39). We hypothesize that proteins gained during in vitro induction are effector molecules of differentiation. Isolation of the controlling genes of such proteins may be useful in determining their role in differentiation.

The idea of inducing malignant B cells to differentiate to a mature non-proliferative state is clinically relevant. However, because of the tumor promoting characteristics of the phorbol esters, these compounds do not have clinical potential. Recently, Pettit et al have isolated a group of compounds known as the bryostatins from marine animals (40,41). Like the phorbol esters, these compounds are PKC activators, however, they lack tumor promoting activity (42) and therefore, can have clinical value as B cell differentiating agents. We have investigated one such agent, bryostatin 1, regarding its ability to induce B cell differentiation in vitro. Our results indicate a comparable activity to that of TPA in inducing differentiation of the Reh ALL cell line (43), non-Hodgkin's lymphomas (44) and CLL (45) (Fig. 4). Like TPA, bryostatin 1 had no effect on Burkitt's lymphoma and multiple myeloma stages.

FUTURE DIRECTIONS

Results of our induction experiments in vitro and those of others form a rational basis for further development of "differentiation therapy" in B cell malignancies. Such a direction is further stimulated by the availability of interesting biological agents like the bryostatins. However, the in vitro data on differentiation has to be verified in a preclinical in vivo model before it can be translated to human trials. Since the monitoring of differentiation induction involves monoclonal antibodies and other markers specific for human

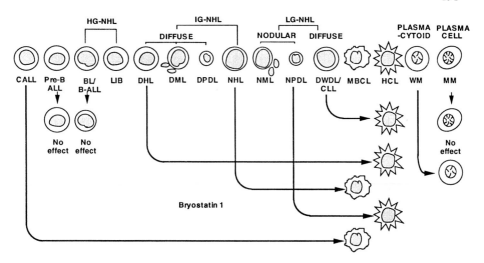

Figure 4: A schematic overview of the differentiating activity of bryostatin 1 on human B cell tumors *in vitro*. Monitoring of differentiation state was determined according to the markers shown in Fig. 1.

cells, the preclinical models should, ideally, be developed with tumor cells of human origin. Such *in vivo* models can then be utilized to validate the *in vitro* findings when the same tumor cells are used in both systems. Based on such reasoning, we have successfully established a xenograft model of human Waldenstrom's macroglobulinemia (WM) in severe combined immune deficient (SCID) mice (32). Such tumors grow as subcutaneous masses at the sites of injection and are easily measurable. We have also demonstrated that such tumors preserve their karyotypes and phenotypes which can be easily assessed by standard flow cytometry techniques similar to those used in the *in vitro* studies. We are currently in the process of developing similar xenograft models for other B cell tumors like the acute lymphoblastic leukemia and non-Hodgkin's lymphoma. Tumor-bearing SCID mice can then be injected with biological

or other agents of potential B cell differentiating activity, like the bryostatins. In addition to tumor measurements as a response criterion, some tumors can be removed and subjected to analysis as outlined in Table 1.

ACKNOWLEDGEMENTS

This work is supported by grants USDHHS R29CA50715
from the National Cancer Institute, the Stephen Brandt
and Dennis Shea Funds for Cancer Research and by the
Ayad Al-Katib Cancer Research Funds of Harper Hospital.
Flow cytometry was performed at the Ben Kasle facility
of the Meyer L. Prentis Comprehensive Cancer Center of
Metropolitan Detroit, supported by USDHHS CA-22453; at
the WSU Department of Immunology and Microbiology Flow
Cytometry facility; and, supported in part by the Center
for Molecular Biology of Wayne State University and the
Department of Immunology and Microbiology. The authors
wish to thank Mark KuKuruga and Eric Van Buren for
assistance in the flow cytometry, Kerry Vistisen for
technical assistance and Cynthia Johnson-Brown for
typing the manuscript.

REFERENCES

1. Market CL: Neoplasia: A disease of cell differen-
    tiation. Cancer Res. 28:1908-1914, 1968.
2. Sachs L: Cell differentiation by bypassing of
    genetic defects in the suppression of malignancy.
    Cancer Res 47:1981-1986, 1987.
3. Anderson KC, Bates MP, Slaughehoupt BL, et al:
    Expression of human B cell-associated antigens on
    leukemias and lymphomas: A model of human B cell
    differentiation. Blood 63:1424-1433, 1984.
4. Hashimi L, Wang CY, Al-Katib A, Koziner B: Cellu-
    lar distribution of a Bcell specific surface anti-
    gen (gp 54) detected by a monoclonal antibody (an-
    ti-BL4). Cancer Res. 46:5431-5437, 1986.
5. Hokland P, Ritz J, Schlossman SF, Nadler LM: Or-
    derly expression of B cell antigens during the in
    vitro differentiation of non-malignant human Pre-B
    cells. J. Immunol. 135:1746-1751, 1985.
6. Loken M, Shah VO, Dattilio KL, Civin CI: Flow
    cytometric analysis of human bone marrow. II. Nor-

mal B lymphocyte development. Blood 70:1316-1324, 1987.

7. Wang CY, Azzo W, Al-Katib A, et al: Preparation and characterization of monoclonal antibodies recognizing three distinct differentiation antigens (BL1, BL2, BL3) on human B lymphocytes. J. Immunol. 133:684-691, 1984.

8. Al-Katib A, Wang CY, Bardales R, Koziner B: Phenotypic characterization on Non-T, Non-B acute lymphoblastic leukemia by a new panel (BL) of monoclonal antibodies. Hematol. Oncol. 3:271-281, 1985.

9. Small TN, Keever CA, Weiner-Fedus S, et al: B-cell differentiation following autologous, conventional, or T-cell depleted bone marrow transplantation: A recapitulation of normal B-cell ontogeny. Blood 76:1647-1656, 1990.

10. Al-Katib A, Koziner B: Leu-10 (HLA-DC/DS) antigen distribution in human leukaemic disorders as detected by a monoclonal antibody: Correlation with HLA-DR expression. Br. J. Haematol. 57:373-382, 1984.

11. Wang CY, Al-Katib A, Lane CL, et al: Induction of HLA-DC/DS (Leu10) antigen expression by human precursor B cell lines. J. Exp. Med. 158:1757-1762, 1983.

12. Cossman J, Neckers LM, Arnold A, Korsmeyer SJ: Induction of differentiation in a case of common acute lymphoblastic leukemia. New Engl. J. Med. 307:1251-1254, 1982.

13. Sugawara I: The immunoglobulin production of human peripheral B lymphocytes induced by phorbol myristate acetate. Cellular Immunol. 72:88-96, 1982.

14. Efremidis AP, Haubenstock H, Holland JF, Bekesi JG: TPA-induced maturation in secretory human B-leukemic cells in vitro: DNA synthesis, antigenic changes, and immunoglobulin secretion. Blood 66:953-960, 1985.

15. Nadler LM, Anderson KC, Marti G, et al: B4, a human lymphocyte associated antigen expressed on normal, mitogen activated and malignant B lymphocytes. J. Immunol. 131:244-250, 1983.

16. Janossy G, Bollum FJ, Bradstock KF, et al: Terminal transferase positive bone marrow cells exhibit the antigen phenotype of common acute lymphoblastic leukemia. J. Immunol. 123:1525-1530, 1979.

17. Patel B, Mohammad RM, Blaustein J, Al-Katib A: Induced expression of a monocytoid B lymphocyte antigen phenotype on the Reh cell line. Am. J. Hematol. 33:153-159, 1990.

18. Liendo C, Danieu L, Al-Katib A, Koziner B: Phenotypic analysis by flow cytometry of surface immunoglobulin light chains and B and T cell antigens in lymph nodes involved with non-Hodgkin's lymphoma. Am. J. Med. 79:445-454, 1985.

19. Vogler LB, Crist WM, Bockman DE, et al: Pre-B-cell leukemia: A new phenotype of childhood lymphoblastic leukemia. N. Engl. J. Med. 298:872-878, 1978.

20. Nadler LM, Stashenko P, Hardy R, et al: Characterization of a human B cell specific antigen (B2) distinct from B1. J. Immunol. 126:1941-1947, 1981.

21. Al-Katib A, Wang CY, Koziner B: Distribution of a new B-cell associated surface antigen (BL7) detected by a monoclonal antibody in human leukemic disorders. Cancer Res. 45:3058-3063, 1985.

22. Schwarting R, Stein H, Wang CY: The monoclonal antibodies α S-HCL1 (α Leu-14) and α S-HCL3 (α Leu-M5) allow the diagnosis of hairy cell leukemia. Blood 65:974-983, 1985.

23. Sheibani K, Burke JS, Swartz WG, et al: Monocytoid B-cell lymphoma: Clinico-pathologic study of 21 cases of a unique type of low-grade lymphoma. Cancer 62:1531-1538, 1988.

24. Traweek ST, Sheibani K, Winberg CD, et al: Monocytoid B-cell lymphoma: Its evaluation and relationship to other low-grade B-cell neoplasms. Blood 73:573-578, 1989.

25. Burke JS, Sheibani K: Hairy cells and monocytoid B lymphocytes: Are they related? Leukemia 1:298-300, 1987.

26. Anderson KC, Park EK, Bates MP, et al: Antigens on human plasma cells identified by monoclonal antibodies. J. Immunol. 130:1132-1138, 1983.

27. Anderson KC, Bates MP, Slaughenhoupt B, et al: A monoclonal antibody with reactivity restricted to normal and neoplastic plasma cells. J. Immunol. 132:3172-3179, 1984.

28. Finerty S, Rickinson AB, Epstein MA, Platts-Mills TAE: Interaction of Epstein-Barr virus with leukemic B cells in vitro. II. Cell line establishment from prolymphocytic leukemia and from Waldenstrom's macroglobulinemia. Int. J. Cancer 30:1-7, 1982.

29. Mohamed AN, Al-Katib A: Establishment and characterization of a human lymphoma cell line (WSU-NHL) with 14;18 translocation. Leuk. Res. 12:833-843, 1988.

30. Mohamed AN, Mohammad RM, Koop BI, Al-Katib A: Establishment and characterization of a new human Burkitt lymphoma cell line (WSU-BL). Cancer 64:1041-1048, 1989.

31. Mohammad RM, Mohamed AN, KuKuruga M, et al: A human B-cell lymphoma line with de novo multidrug resistance phenotype. Cancer 69:1468-1474, 1992.

32. Al-Katib A, Mohammad R, Hamdan M, et al: Propagation of Waldenstrom's macroglobulinemia cells in vitro and in severe combined immune deficient mice: Utility as a preclinical drug screening model. Blood 81:3034-3042, 1993.

33. Mohammad RM, Smith MR, Mohamed AN, Al-Katib A: A unique EBV-negative low grade lymphoma line (WSU-FSCCL) exhibiting both t(14;18) and t(8,11) chromosomal translocations. Cancer Genet. Cytogenet. 70:62-67, 1993.

34. Hamdan M, Mohammad RM, Mohamed A, Al-Katib A: Development of a xenograft model for diffuse large cell human lymphoma (DLCL) in SCID mice. Proc., AACR, abst #349, 1993.

35. Rosenberg SA, Bernard CW, Brown BW Jr, et al: National Cancer Institute sponsored study of classification of non-Hodgkin's lymphomas: Summary and description of working formulation for clinical usage. Cancer 49:2112-2135, 1982.

36. Totterman TH, Nilsson K, Sundstrom C: Phorbol ester-induced differentiation of chronic lymphocytic leukemia cells. Nature 288:176-178, 1980.

37. Al-Katib A, Wang CY, McKenzie S, et al: Phorbol ester-induced hairy cell features on chronic lymphocytic leukemia cells in vitro. Am. J. Hematol. 40:264-269, 1992.

38. Mohamed AN, Nakeff A, Mohammad RM, et al: Modulation of c-myc oncogene expression by phorbol ester and interferon-gamma: Appraisal by flow cytometry. Oncogene 3:429-435, 1988.

39. Al-Katib A, Mohammad RM, Mohamed AN, et al: Conversion of high grade lymphoma tumor cell line to intermediate grade with TPA and bryostatin 1 as determined by polypeptide analysis on 2D gel electrophoresis. Hematol. Oncol. 8:81-89, 1990.

40. Pettit GR, Day JF, Hartwell JL, Wood HB: Antineoplastic components of marine animals. Nature 227:962-963, 1970.

41. Pettit GR, Herald SL, Doubek DL, et al: Isolation and structure of bryostatin 1. J. Am. Chem. Soc. 104:6846-6848, 1982.

42. Hennings H, Blumberg PM, Pettit GR, et al: Bryostatin 1, an activator of protein kinase C, inhibits tumor promotion by phorbol ester in SENCAR mouse skin. Carcinogenesis 8:1343-1346, 1987.

43. Al-Katib A, Mohammad RM, Khan K, et al: Bryostatin 1-induced modulation of the acute lymphoblastic leukemia cell line Reh. J. Immunother. 14:33-42, 1993.

44. Mohammad RM, Al-Katib A, Pettit GR, Sensenbrenner LL: Differential effects of bryostatin 1 on human non-Hodgkin's lymphoma cell lines. Leuk. Res. 17:1-8, 1993.

45. Al-Katib A, Mohammad RM, Dan M, et al: Bryostatin 1-induced hairy cell features on chronic lymphocytic leukemia cells in vitro. Exp. Hematol. 21:61-65, 1993.

13

PROBING IMMUNE RESPONSES: THE ROLE OF INTRACELLULAR
GLUTATHIONE

Mario Roederer and Leonard A. Herzenberg

INTRODUCTION

The power of flow cytometry is evidenced by its
ability to measure a variety of parameters on a cell-by-
cell basis. The utility of flow cytometric measurements
of these parameters becomes evident when it is understood that, for a population of cells, the <u>distribution</u>
of these parameters can be determined. Thus, the heterogeneity within cell populations can be determined.
This obviates the need for the assumption that an average value obtained for a population of cells (as obtained using a bulk analysis) is representative for each
cell within that population.

We have been studying the role of both inflammatory
and oxidative stresses that are present during the
progression of the HIV infection. These studies were
prompted by the observations that (1) glutathione (GSH)
levels in several tissues from infected individuals was
decreased; and, (2) levels of inflammatory cytokines
(especially, tumor necrosis factor alpha (TNF), IL-1,
and IL-6) are present in the sera from infected individuals. We hypothesized that in fact these two were directly and perhaps causally related (1).

In order to study the role that intracellular GSH
might play in the HIV infection, we adapted a flow
cytometric assay for the determination of intracellular
GSH levels in order to quantitate the cell-associated

GSH for subsets of peripheral blood mononuclear cells (PBMC). This is a relatively simple assay, relying on the enzyme-catalyzed reaction of monochlorobimane (MCB) with GSH, to form a fluorescent adduct which is trapped within cells. The fluorescence is measured by UV excitation, and is proportional to the cellular GSH content.

Combining the GSH assay with surface immunofluorescence staining demonstrated that each subset within PBMC as a whole the distribution becomes far less complex when subsets are examined individually. Thus, B cells, monocytes, and NK cells have relatively homogeneous levels of GSH. Interestingly, T cells can be divided into two classes of cells, containing high and low levels of GSH, respectively (see below) (2).

By applying this assay to PBMC obtained from 26 healthy individuals, we find that the level of GSH varies only a small amount for each subset taken individually. Thus, for example, the variation in B cell GSH from individual to individual is less than 10 percent. This strongly suggests that the level of GSH, while different for each cell type within PBMC, is highly regulated by the cells (2).

Thus, it was surprising to find such a strong heterogeneity in the GSH levels in CD4 and CD8 T cells. Such a heterogeneity is not found within other subpopulations in man, and not found in any subpopulations in mice. Detailed examination of the phenotypes of these cells revealed that some T cell markers are preferentially found on the high-GSH subsets (and some on the low-GSH subsets). Perhaps the most striking is the observation that virtually all of the high-GSH T cells are CD45RA[+] ("virgin" T cells) (Fig. 1). In addition, the high-GSH T cells generally do not display any of the markers associated with activation, suggesting that these cells may perhaps be newly-arisen and unstimulated T cells.

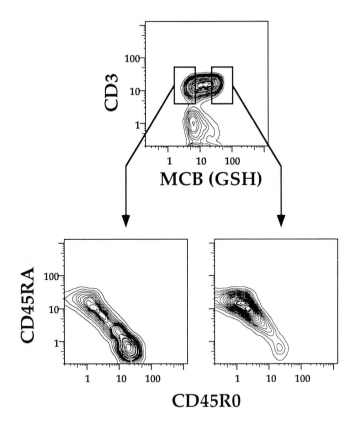

Figure 1. High-GSH T cells are predominantly CD45RA+. PBMC from a normal individual were simultaneously stained for GSH, CD3, CD45RA, and CD45RO. The high-GSH T cells and low-GSH T cells were selected by software gating (top panel); the CD45RA and CD45RO distributions for these cells is shown (bottom panels). The high-GSH T cells are almost exclusively CD45RA+; the low-GSH T cells are of both CD45RA+ and CD45RO+ phenotypes.

This hypothesis is strengthened by the observation that heterogeneous GSH distributions are seen in fetal thymic tissues, but only in the mature "single positives" (Fig. 2). The distribution of GSH is quite different in the phenotypically-distinguished thymocytes. Immature, CD3-dim, CD4 and CD8 double positive thymocytes have a uniform, low GSH distribution. However, the CD3-bright, CD4 or CD8 single-positive thymocytes (which have undergone selection) show a much broader distribu-

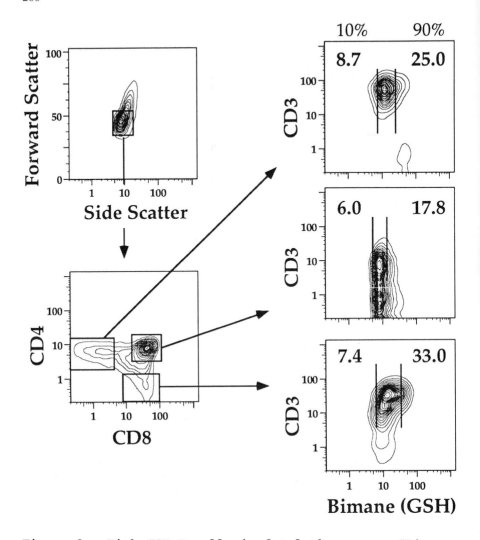

Figure 2. High-GSH T cells in fetal thymus are CD4 or CD8 single-positive. Thymocytes from a 22-week fetal thymus were simultaneously stained for GSH, CD3, CD4, and CD8. Software analysis (gating) was used to select lymphocytes (top left panel: size gate); the distribution of CD4 and CD8 on these cells is shown in the bottom left panel. From these cells, the CD3 vs GSH distribution (right panels) is shown for CD4 single positives (top), CD4 CD8 double positives (middle), and CD8 single positives (bottom). Within each panel is the MCB fluorescence (linearly related to GSH level) for the 10th and 90th percentile of the distribution (also marked by vertical bars). Note that the double positives are narrowly distributed, as compared to the single-positives.

tion of GSH, like T cells in the periphery.

Groups led by Droge and Crystal have demonstrated that there is a GSH defeciency in HIV-infected individuals (3,4). These results prompted us to examine the levels of GSH in PBMC from these people. By measuring GSH in PBMC subsets, we confirmed the general decrease observed previously (5). However, we found that there is considerable differences amongst subsets within PBMC. While there is a disregulation of intracellular GSH in all PBMC subsets, T cells were most dramatically affected: Both CD4 and CD8 T cells have (on average) 60% of the intracellular GSH from uninfected individuals. Careful analysis of the distribution of GSH in these cells revealed that there was a specific loss from the periphery of the high-GSH T cells, whereas the low-GSH T cells did not appear to lose GSH (2,5).

The observations that intracellular GSH levels are depleted in T cells from HIV-infected individuals raises the question: What effect does this depletion have on functionality? The role of intracellular GSH in T cell function has been studied for many years, and its importance has been established.

Wedner and colleagues first showed that depletion of GSH inhibited T cell proliferation (6). In several studies, Fidelus et al also showed the critical need for adequate intracellular GSH levels for T cell proliferation (7-9). Reductions of GSH by 10-40% in T cells completely inhibited T cell activation (10). Messina et al showed that a 40% decrease in intracellular GSH inhibited cell cycle progression in PBMC. However, even a 90% decrease in intracellular GSH did not inhibit the stimulated expression of IL-2 receptor or secretion of IL-2 (11). This is not surprising, in view of the dependence of IL-2 receptor expression on activation of NF-κB (12-14): This activation is more likely under conditions of GSH-depletion (see below). Gmunder et al confirmed this result (15), and further suggested that some processes are GSH-dependent, while others are cysteine-dependent (16). There seemed to be no linear correlation between the degree of GSH depletion and the inhibition observed (10).

While depletion of GSH leads to inhibition of some T cell functions (but increased sensitivity to inflammatory cytokines), supplementation can augment other functions, both in vitro and in vivo. Exogenously-added GSH augments lymphocyte proliferation in response to lectin (17,18). Oxothiazolidine 4-carboxylate (OTC), a cysteine precursor which increases GSH levels, acts synergistically with Concanavalin A to stimulate T cells (7). Finally, Droge et al showed that increasing previously lowered GSH levels in mice augments the activation of cytolytic T cells, demonstrating the importance of GSH levels in vivo (19).

We have used a number of different model systems to study the role of intracellular thiols (and exogenously-added thiols) in the regulation of inflammatory stimulations; specifically, those resulting in increased expression of HIV. We and others have previously shown that the TNF-mediated stimulation of HIV could be blocked by addition of N-acetylcysteine (NAC), a cysteine-prodrug which can replenish GSH levels in vivo (20-23). One of the models we commonly use is the acute infection of PBMC by laboratory strains of HIV. We have used this model to demonstrate that 1 mM NAC can completely block the TNF-mediated stimulation of HIV replication in PHA-blasted PBMC (20). At higher concentrations, NAC blocks the replication of HIV in cultures without exogenous stimulation (albeit, blasted PBMC cultures produce many cytokines capable of stimulating HIV; we hypothesize that NAC interferes with the signal transduction for at least some of these molecules as it does for TNF).

To address the question of what effect depleted GSH levels might have on this model, we used a specific inhibitor of GSH synthesis, buthionine sulfoximine (BSO). Incubation of blasted PBMC for 3 days in the presence of 100 μM BSO depleted 40% of intracellular GSH. This is a similar depletion to that found in T cells from HIV-infected individuals. Both untreated and BSO-treated cultures were then acutely infected with HIV; TNF and PMA and/or NAC were then added to the cultures. Viral replication (as measured by production of the viral core protein p24 in the supernatant) was assessed after 3 days.

As shown in Fig. 3, simply decreasing GSH levels resulted in an increase in viral replication in the PBMC. It is unlikely that this is a direct effect; rather, it is likely that lowering GSH levels makes the cells more susceptible to stimulation by inflammatory cytokines (which are present in these cultures). This is substantiated by the observation that 1 mM NAC inhibited replication of HIV, even in the cultures treated with BSO. Note that since the BSO treatment was continuous, this rules out that GSH synthesis is required for the inhibition by NAC.

These results demonstrate that there may be a profound effect of lowered GSH on the functionality of T cells in infected individuals. Besides the known dependence of T cell function on intracellular GSH levels, it is also possible that the lowered levels can lead to increased viral replication. This has led us and others to suggest that GSH-replenishment therapy be an adjunct in the treatment of AIDS.

What is the mechanism by which NAC inhibits the signal transduction by TNF and PMA? Trying to determine the precise effect of an inhibitor is always fraught with difficulty; however, progress is slowly being made. The inhibition is definitely in the signal transduction

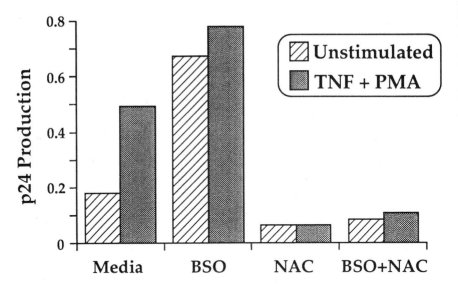

Figure 3. Depletion of GSH in PBMC results in increased HIV replication. PBMC were blasted with PHA for 3 d as previously described (21). Cultures were then split into 2 aliquots, and maintained for 3 d in media supplemented with IL-2 (without PHA). BSO was added to one aliquot to a final concentration of 100 μM. A small portion of each aliquot was used at this point to determine the intracellular GSH level by FACS; the BSO-treated cultures showed a 40% reduction of GSH (data not shown). The aliquots were then acutely infected with HIV and stimulated in the presence or absence of 1 mM NAC and 100 μM BSO for 3 d as described (21). Viral replication (as measured by the production of p24) was enhanced by GSH depletion (BSO-treated cultures), and by TNF and PMA stimulation. Together, GSH depletion and stimulation resulted in greater stimulation than by either alone. 1 mM NAC completely inhibited the enhanced replication either by GSH depletion or by TNF and PMA.

pathway, in that activation of NF-κB is blocked in the presence of NAC (23,24). This DNA enhancer-binding protein is normally found inactive in the cytoplasm; its activation occurs rapidly after stimulation and is considered an "early" event during inflammatory stimulation. NF-κB activation is a necessary prerequisite for the stimulation of HIV by TNF (and PMA); it is suffi-

cient for stimulation of HIV. Baeuerle has suggested
that the production of oxidative intermediates is a
necessary step in the activation of NF-κB, and that
thiols interfere with this process (25). However, it is
still unclear as to what point in the signal transduction pathway is directly affected.

In order to study the mechanism of HIV regulation,
we use a model in which the HIV LTR (promoter/enhancer
regions for HIV) is fused to a reporter gene, lacZ.
Measurements of the gene product ß-galactosidase reveal
the transcriptional activity of the HIV LTR. This model
reproduces known regulation of HIV quite well (20).
Furthermore, assays to quantitate the level of the gene
product can be conveniently performed and quantitated in
the same 96-well tray format as the stimulations, which
results in the ability to do thousands of independent
assays in a single day.

Careful kinetic analysis demonstrates that the NAC-
sensitive step in signal transduction occurs about 20-40
min after stimulation (Fig. 4). Cells stably transfec-
ted with the reporter construct were stimulated with
TNF, PMA, or both; at various time points with respect
to the stimulation, NAC was added to the cultures. Six
hours later, the total ß-galactosidase activity was
measured. As shown in Fig. 4, NAC is equally effective
when added 20 min after TNF as when it is added concurrently. This suggests that the sensitive step in signal
transduction does not occur for at least 20 min. NAC
added after this point only blocks part of the signal.

PMA stimulation is more sensitive to NAC (in terms
of concentration) than TNF (20). Furthermore, the NAC-
sensitive step in signal transduction is delayed for TNF
stimulation compared to PMA stimulation. Note that, as
expected, the first NAC-sensitive step for the synergistic stimulation is at the same time as the earliest for
each stimulation alone. These observations suggest that

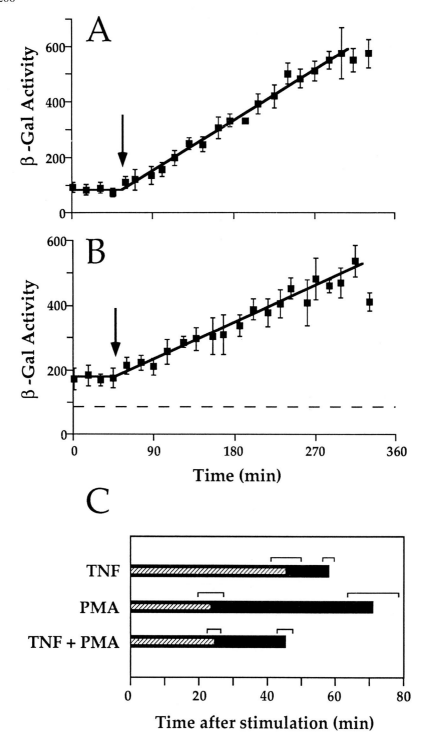

Figure 4 (see previous page). The NAC-sensitive step in signal transduction precedes stimulated protein synthesis by 20-40 min. 293.27.2 cells, which contain a stably-integrated HIV LTR-lacZ gene fusion, were stimulated with TNF and/or PMA in the presence or absence of 30 mM NAC, as previously described (20). (A,B) ß-gal activity in cultures was measured by the MUG assay (26). Error bars represent ±1 sd for 8 separate wells for each condition. (A) TNF stimulates ß-gal production in a time-dependent fashion. Over a short time period (less than 5 h), the increase in activity is linear. Extrapolation shows that the appearance of new protein in response to TNF occurs approximately 60 min after addition of the cytokine. (B) The inhibition by NAC precedes protein synthesis. All samples were stimulated with TNF at time t=0 min. Thereafter, NAC was added to separate cultures at 15 min intervals; the total ß-gal activity was measured 6 h after the initial stimulation with TNF. Thus, NAC can be added as long as 30-45 min after TNF and still achieve maximal inhibition. Thereafter, NAC can only inhibit a fraction of the stimulated protein synthesis. (C) Experiments such as those in (A) and (B) were repeated for stimulations with PMA and TNF+PMA. The filled bar shows the average time after stimulation before protein synthesis begins. The hatched bar is the earliest time point at which NAC ceased to fully inhibit stimulation. Error bars represent ±1 sd for three independent experiments. Each condition has a characteristic length of time for signal transduction to be completed; the NAC-sensitive step in each pathway is also at different times after addition of the stimulus.

there are distinct mechanisms by which NAC inhibits signal transduction pathways - and thus that the redox regulation imposed by intracellular GSH may differentially affect a wide variety of cellular processes.

These results, together with the previously published observations on the role of intracellular GSH in T cell function, suggest that the redox regulation acts oppositely on inflammatory stimulations and non-inflammatory stimulations. Thus, in general, TNF-mediated stimulations are inhibited by agents which replenish GSH, but are exacerbated under conditions where GSH is depleted. On the other hand, antigen-mediated or lectin-mediated stimulation and mitosis is inhibited by GSH

depletion, but potentiated by GSH-supplementation. This
regulation is of clinical relevance in AIDS, in that
HIV-infected individuals show decreased GSH levels
together with evidence of chronic inflammatory stimulation (1).

ACKNOWLEDGEMENTS

We thank Paul Raju for expert technical support.
Mario Roederer is a Senior Fellow of the Leukemia Society of America. This work was supported in part by NIH
grant CA-42509 to Leonard A. Herzenberg.

REFERENCES

1. Roederer M, Ela SW, Staal FJT, et al: N-acetylcys-
    teine: A new approach to anti-HIV therapy. AIDS
    Res. Human Retr. 8:209-217, 1992.
2. Roederer M, Staal FJT, Osada H, et al: CD4 and CD8
    T cells with high intracellular glutathione levels
    are selectively lost as the HIV infection progress-
    es. Intl. Immunol. 3:933-937, 1991.
3. Eck H-P, Gmunder H, Hartmann M, et al: Low Concen-
    trations of acid-soluble thiol (cysteine) in the
    blood plasma of HIV-1-infected patients. Biol.
    Chem. Hoppe-Seyler 370:101-108, 1989.
4. Buhl R, Holroyd KJ, Mastrangeli A, et al: Systemic
    glutathione deficiency in asymptomatic HIV-seropos-
    itive individuals. Lancet ii:1294-1298, 1989.
5. Staal FJT, Roederer M, Israelski DM, et al: Intra-
    cellular glutathione levels in T cell subsets de-
    crease in HIV infected individuals. AIDS Res.
    Human Retr. 8:311-318, 1992.
6. Fischman CH, Udey MC, Kurtz M, Wedner HJ: Inhi-
    bition of lectin-induced lymphocyte activation by
    2-cyclohexene-1-one: Decreased intracellular gluta-
    thione inhibits an early event in the activation
    sequence. J. Immunol. 127:2257-2262, 1981.
7. Fidelus RK, Tsan M-F: Enhancement of intracellular
    glutathione promotes lymphocyte activation by mito-
    gen. Cell. Immunol. 97:155-163, 1986.
8. Fidelus RK, Ginouves P, Lawrence D, Tsan M-F:
    Modulation of intracellular glutathione concentra-
    tions alters lymphocyte activation and prolifera-
    tion. Exp. Cell Res. 170:269-275, 1987.
9. Fidelus RK: The generation of oxygen radicals: A
    positive signal for lymphocyte activation. Cell.
    Immunol. 113:175-182, 1988.
10. Suthanthiran M, Anderson ME, Sharma VK, Meister A:

Glutathione regulates activation-dependent DNA synthesis in highly purified normal human T lymphocytes stimulated via the CD2 and CD3 antigens. Proc. Natl. Acad. Sci. USA 87:3343-3347, 1990.

11. Messina JP, Lawrence DA: Cell cycle progression of glutathione-depleted human peripheral blood mononuclear cells is inhibited at S phase. J. Immunol. 143:1974-1981, 1989.

12. Ballard DW, Bohnlein E, Lowenthal JW, et al: HTLV-I Tax induces cellular proteins that activate the kappa B element in the IL-2 receptor alpha gene. Science 241:1652-1655, 1988.

13. Lowenthal JW, Ballard DW, Bohnlein E, Greene WC: Tumor necrosis factor alpha induces proteins that bind specifically to kappaB-like enhancer elements and regulate interleukin 2 receptor alpha-chain gene expression in primary human T lymphocytes. Proc. Natl. Acad. Sci. USA 86:2331-2335, 1989.

14. Shibuya H, Yoneyama M, Taniguchi T: Involvement of a common transcription factor in the regulated expression of IL-2 and IL-2 receptor genes. Intl. Immunol. 1:43-49, 1989.

15. Gmunder H, Roth S, Eck H-P, et al: Interleukin-2 messenger RNA expression, lymphokine production and DNA synthesis in glutathione-depleted T-cells. Cell. Immunol. 130:520-528, 1990.

16. Gmunder H, Eck H-P, Benninghoff B, et al: Macrophages regulate intracellular glutathione levels of lymphocytes. Evidence for an immunoregulatory role of cysteine. Cell. Immunol. 129:32-46, 1990.

17. Hamilos DL, Wedner HJ: The role of glutathione in lymphocyte activation. I. Comparison of inhibitory effects of buthionine sulfoximine and 2-cyclohexene-1-one by nuclear size transformation. J. Immunol. 135:2740-2747, 1985.

18. Hamilos DL, Zelarney P, Mascali JJ: Lymphocyte proliferation in glutathione-depleted lymphocytes: Direct relationship between glutathione availability and the proliferative response. Immunopharmacology 18:223-235, 1989.

19. Droge W, Pottmeyer-Gerber C, Schmidt H, Nick S: Glutathione augments the activation of cytotoxic T lymphocytes in vivo. Immunobiol. 172:151-156, 1986.

20. Roederer M, Staal FJT, Raju PA, et al: Cytokine-stimulated HIV replication is inhibited by N-acetylcysteine. Proc. Natl. Acad. Sci. USA 87:4884-4888, 1990.

21. Roederer M, Raju PA, Staal FJT, et al: N-Acetylcysteine inhibits latent HIV expression in chronically-infected cells. AIDS Res. Human Retr.7:491-495, 1991.

22. Kalebic T, Kinter A, Poli G, et al: Suppression of HIV expression in chronically infected monocytic cells by glutathione, glutathione ester, and N-

acetylcysteine. Proc. Natl. Acad. Sci. USA 88:986-990, 1991.

23. Mihm S, Ennen J, Pessara U, et al: Inhibition of HIV-1 replication and NF-kappaB activity by cysteine and cysteine derivatives. AIDS 5:497-503, 1991.

24. Staal FJT, Roederer M, Herzenberg LA, Herzenberg LA: Intracellular thiols regulate activation of nuclear factor κB and transcription of human immunodeficiency virus. Proc. Natl. Acad. Sci. USA 87:9943-9947, 1990.

25. Schreck R, Rieber P, Baeuerle PA: Reactive oxygen intermediates as apparently widely used messengers in the activation of the NF-kappaB transcription factor and HIV-1. EMBO J. 10:2247-2258, 1991.

26. Roederer M, Fiering SN, Herzenberg LA: FACS-Gal: Flow cytometric analysis and sorting of cells expressing reporter gene constructs. Methods: A Companion to Methods in Enzymology 2:248-260, 1991.

14

DNA TOPOISOMERASE II POISONS AND THE CELL CYCLE

P.J. Smith

INTRODUCTION

Nuclear DNA, a target for many anticancer drugs,
does not exist inside a human cell as an extended,
relaxed and unrestrained molecule. Rather, in its pro-
tein-associated form of chromatin it is highly organized
with order being imposed by specific DNA-protein interactions. The imposition of such restraints on the free
movement of DNA molecules within the nucleus means that
any topological problems must be actively resolved by
the topoisomerase enzymes (1). The major type II enzyme, DNA topoisomerase II, is strategically located at
the base of chromatin loop domains to carry out its
housekeeping functions and operates by forming temporary
gates in double stranded DNA (Fig. 1) through which an
intact helix can pass. Several classes of antitumor
drugs are now recognized as topoisomerase poisons because of their ability to trap the enzyme gates on DNA
in the form of stabilized cleavable complexes (Fig. 1).
The term "cleavable" refers to the ability of strong
protein denaturants to reveal that the trapped enzyme
sequesters a DNA double strand break within the complex
(Fig. 1) (2). Trapped complexes are thought to give
rise to potentially lethal lesions, probably unrepaired
double strand breaks. Importantly, the intrinsic sensitivity of actively proliferating tumor cells to topoisomerase II poisons appears to be a function of the

212

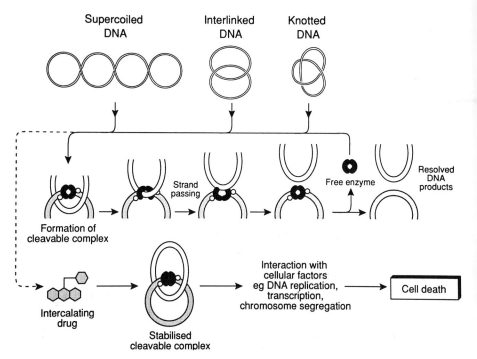

Figure 1. The resolution of topological problems with DNA by the nuclear enzyme DNA topoisomerase II. The enzyme forms a protein gate (cleavable complex) in DNA through which an intact helix can pass. The cleavable complexes become stabilized (poisoned) in the presence of some antitumor drugs. Reprinted with permission from ref. 3.

availability of the target enzyme (2).

Specific classes of DNA intercalating and non-intercalating agents can act as topoisomerase II poisons and examples of such agents are shown in Fig. 2. The anilinoacridine derivative, m-AMSA, binds to DNA through weak, reversible intercalation and has high activity against acute leukemia and malignant lymphoma. The bis(alkylamino) anthraquinone, mitoxantrone, is effective in the treatment of acute leukemia, lymphoma and breast cancer. Although binding of mitoxantrone to nucleic acids is by intercalation, the long alkyl side-chains bind electrostatically with the anionic exterior of the helix and prevent complete intercalation (4).

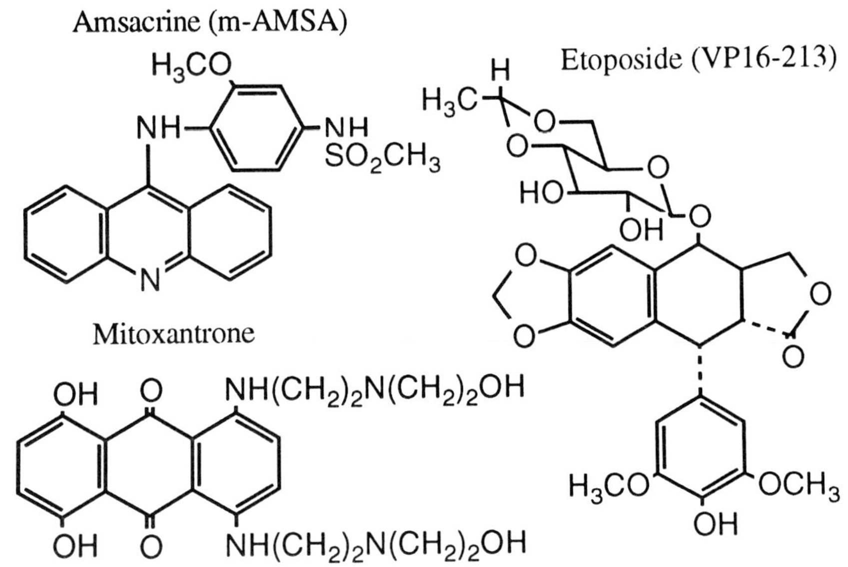

Figure 2. Examples of antitumor topoisomerase poisons

The most widely used class of non-intercalative antitumor topoisomerase II poisons is the epipodophyllotoxins including etoposide (VP16-213) and teniposide (VM-26). VP16-213 is currently used in the treatment of acute nonlymphocytic leukemia, Hodgkin's disease, teratoma of the testis, and in combination with Adriamycin against various lymphomas. The effects of epipodophyllotoxins on cultured cells are similar to those of the intercalative agents described above, and include the induction of DNA strand breakage (5), DNA-protein crosslinking (6) and inhibition of DNA synthesis (7). There is convincing evidence that DNA topoisomerase II is the primary cellular target for both VP16-213 and VM-26 (8) although additional factors may determine the cytotoxic action of DNA intercalating agents.

This brief overview attempts to provide examples of how DNA topoisomerase II can define cellular sensitivity to drug-induced cell cycle arrest by virtue of its intranuclear availability and the mode of enzyme trapping on DNA.

MATERIALS AND METHODS

Cell Culture Drug Treatments

The established small cell lung cancer (SCLC) cell line, NCI-H69/P (designated H69; kindly provided by Dr. Peter Twentyman, MRC Centre, Cambridge) was grown under standard suspension culture conditions in RPMI medium supplemented with 10% fetal calf serum, glutamine and antibiotics. SCLC cells grew as small aggregates of less than 50 cells and large spheroid formation was prevented by repetitive pipetting of the culture medium every 2-3 days. The SV40 transformed human fibroblast cell lines MRC5CV1 (normal donor) and AT5B1VA (ataxia-telangiectasia homozygote donor) were grown as asynchronous cultures in Eagle's minimum essential medium supplemented with 10% fetal calf serum, 1 mM glutamine and antibiotics and incubated at $37^{\circ}C$ in an atmosphere of 5% CO_2 in air. The drug VP16-213 (Vepesid; Bristol Myers Pharmaceuticals, Syracuse, NY) was provided as a 34 mM stock solution. Mitoxantrone (Novantrone; a gift from Lederle Laboratories, Gosport, UK) was stored at $-20^{\circ}C$ as a 2 mM aqueous stock. m-AMSA (amsacrine, Park Davis, Eastleigh, UK) was stored at $-20^{\circ}C$ as a 10 mM stock in dimethylsulphoxide.

Single Cell Analysis of Nuclear DNA Topoisomerase II Content

Samples of approximately 1×10^6 cells were washed in buffer and gently permeabilized using the technique described previously (9). Permeabilized cells were fixed in 50% methanol (v/v) and agitated for 30 min at $4^{\circ}C$. Fixed cells were washed once in PBS and resuspended in 20 μl anti-topoisomerase II (p170 form) affinity purified antibody (kindly supplied by Cambridge Research Biochemicals, ICI, Alderley Park, UK; 1:4 dilution) and held for 1 h at room temperature. Antibody treated samples were washed once in PBS and resuspended in 20 μl

FITC-conjugated sheep anti-rabbit IgG (1:100 dilution; Sigma Chemicals, whole molecule) and held at room temperature for 30 min. Finally, samples were pelleted and resuspended in PBS containing ribonuclease A and propidium iodide (5 μg/ml) to stain nuclear DNA. Control samples were processed as above but without the first anti-topoisomerase II antibody treatment. The analysis of samples by flow cytometry has been described previously (10) providing dual fluorescence analysis of cell populations gated for the elimination of debris and cell clumps. The right-angle fluorescence (RF) parameters monitor DNA content (630 nm RF) and second antibody binding (530 nm RF).

Cell Cycle Analysis and Detection of Mitotic Subpopulations

Cells were stained with ethidium bromide (50 μg/ml) plus 0.125% Triton X-100 and ribonuclease (0.5 μg/ml) for 10 min prior to analysis. DNA fluorescence distributions were analyzed by a computer using a cell cycle phase fitting program which assumes normal distributions for G1 and G2/M phase populations. A probability function is calculated for the S phase distribution based upon the means and standard deviations of the G1 and G2/M phase. For stathmokinetic experiments, etoposide-treated and control cultures were exposed to colcemid (60 ng/ml) in order to induce mitotic arrest and low scatter mitotic populations analyzed as described previously (11).

RESULTS

Overproduction of Topoisomerase II and Increased Drug Sensitivity

Two human transformed fibroblast cell lines have been described which differ consistently in the level of DNA topoisomerase II per cell (10). The overproducing line, AT5BIVA is significantly more sensitive to the cytotoxic actions of the topoisomerase II poisons, VP16, m-AMSA and mitoxantrone compared to a normal cell line MRC5CVI. The intrinsic sensitivity of a cell to drug-induced cell cycle arrest is also dependent upon target enzyme availability. Fig. 3 shows a comparison of the effects of VP16-213 on cell cycle progression of the two cell lines. Clearly, both the dose-dependent arrest of cells in S phase and the time-dependent trapping of cells in G2 occur at lower drug concentrations in the overproducing cell line compared to the normal control.

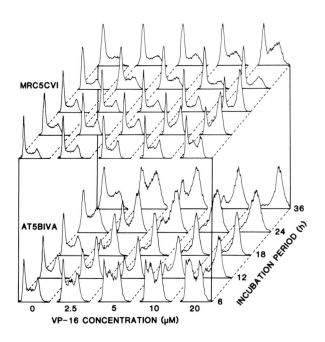

Figure 3. DNA content/frequency distribution profiles for two transformed human fibroblast lines as a function of time after treatment with VP16-213 (30 min exposure). Reprinted with permission from ref. 12.

Differences Between DNA Intercalators in the Mode of Topoisomerase II Poisoning

Despite sharing a similar capacity for the intranuclear trapping of topoisomerase II, the two intercalat-

ors m-AMSA and mitoxantrone differ in their cytotoxic
potency. Analyses of clonogenic survival data for
MRC5CVI cells give D_O values of 121.8 nM for mitoxantrone compared with 2.9 μM for m-AMSA indicating that the
human cell line is 24-fold more sensitive to mitoxantrone than to m-AMSA.

Cell cycle analyses of cultures treated with drugs
for 1 h (Fig. 4) show that although both agents induce
G2/M delay at concentrations which are cytotoxic, the
kinetics of delay differ. The degree of G2 delay induced by m-AMSA was maximal 12 h after removal of drug
and minimal by 24 h, while that induced by mitoxantrone
increased with time up to at least 48 h. The slow recruitment of cells into G2 delay can be attributed to a
long-term depression of DNA synthesis by mitoxantrone
compared with the relatively rapid recovery of DNA
synthesis patterns in m-AMSA-treated cells (13). Fig. 5
shows that although there is a rapid loss of m-AMSA-
induced cleavable complexes (80% of the complexes being

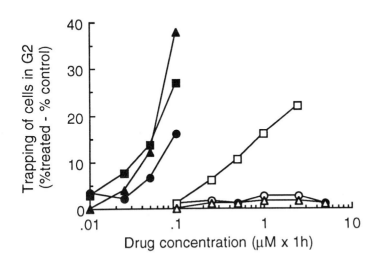

Figure 4. Toxicity of 1 h exposure to mitoxantrone
(closed symbols) or m-AMSA (open symbols) determined by
the retention of cells in G2/M phase of the cell cycle.
Data derived from ref. 13. Post-treatment incubation
periods were: O, ●, 6 h; □, ■, 12 h; △, ▲, 24 h.

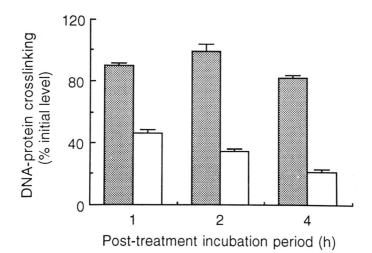

Figure 5. Persistence of topoisomerase II trapping (monitored as drug-induced DNAprotein crosslinking) in MRC5CV1 cells during a 4 h post-treatment incubation period following 1 h exposure to 1 μM mitoxantrone (shaded) or 2.5 μM m-AMSA (open). Data derived from ref. 13.

resolved within 4 h of removal of the drug), mitoxantr-one-induced complexes persist at a high level. Persistence of damage is likely to be a critical factor in determining recovery from cell cycle delay.

Cell Cycle Regulation of Topoisomerase II in SCLC Cells: Implications for Therapy

In accordance with its putative role in the decatenation of DNA replication products, topoisomerase II content increases as cells progress through S phase towards mitosis (14). Fig. 6 demonstrates the cell cycle distribution of enzyme levels in asynchronous SCLC cells and in cultures which have been perturbed by the presence of a mitosis blocking agent, colcemid.

Clearly, cells accumulate in later stages of the cell cycle in the presence of colcemid but retain the levels of enzyme appropriate for their position in the cell cycle. Consequently, target enzyme availability is effectively enhanced in such perturbed populations.

Low doses of cytotoxic drugs are more likely to produce quasi-syncronization of tumor populations rather than immediate cytostasis and cell death, raising the possibility that chronic low dose therapy may have dual effects. On one hand, low dose therapy would reduce tumor proliferation rate while on the other hand, enhancing the intrinsic sensitivity of a tumor population by elevating the availability of a cell cycle regulated target enzyme such as topoisomerase II. Indeed, in the case of SCLC, VP16-213 has been used in combination therapy with <u>in vitro</u> and <u>in vivo</u> studies indicating

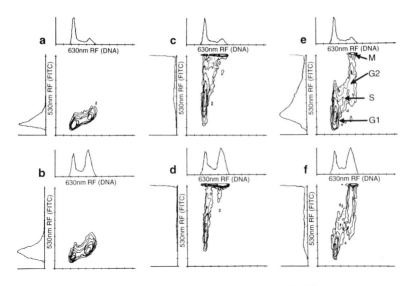

Figure 6 a-f. Contour plots of dual fluorescence analyses vs frequency for control (panels a, c and e) and colcemid-treated (panels b, d and f) H69 SCLC cells. The right-angle fluorescence (RF) parameters monitor DNA content (630 nm RF) and binding of an FITC-labelled second antibody (530 nm RF) directed at a bound anti-topoisomerase II antibody. In all cases cell populations were gated to exclude cell debris and clumps. The projected monodimensional frequency distribution for each axis parameter is scaled to the maximum channel within the data set. Panels: a and b, no first antibody treatment; c and d, with first antibody treatment; e and f, with first antibody treatment and gain settings for 530 nm RF detection reduced by 50%.

that etoposide is a schedule-dependent cytotoxic drug. The importance of longer duration of exposure to etoposide was indicated by a recent study on previously untreated SCLC patients (15), with the implication that continuous exposure to "low" VP16-213 concentrations is required for optimal cytotoxic effect. Using the mitotic spindle inhibitor colcemid, it is possible to investigate the cell cycle perturbations caused by low doses of VP16-213 without the complications of cell division and resupply of G1 phase. Bivariate analysis of right-angle light scatter (488 nm) characteristics and DNA content (530 nm) of cells permits the identification of a low scatter population (LSP) in G2 representing cells entering mitosis (11). Gating three regions (G1+S, G2 and LSP) on contour plots gives an estimation of the proportion of drug-treated cells moving through the cell cycle and the number attempting mitosis (Fig. 7).

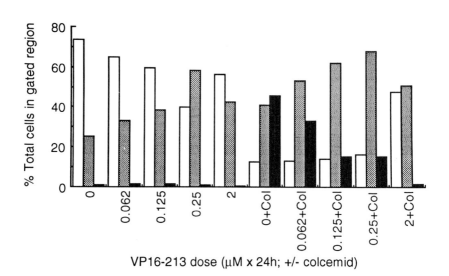

Figure 7. Stathmokinetic analysis of VP16-213-induced cell cycle delay in H69 cells. Open columns, G1 + S phase; shaded columns, G2; closed columns, LSP (mitotic subset).

The data in Fig. 7 show that although 24 h exposure to VP16-213 alone results in the accumulation of cells in G2, as a function of dose, balanced by an emptying of G1+S phase with no evidence of trapping of cells in mitosis (low scatter population, LSP). In the presence of colcemid, the percentage of cells attempting mitosis decreases as a function of dose and is approximately 1% at 2 μM VP16-213 (corresponding to the active plasma concentration of the drug). High dose (2 μM) VP16-213 induces a significant delay in the delivery of cells to G2 and a complete block to G2 exit. Thus, it appears that at low doses, over the first 24 h, cells are progressing through the cell cycle relatively normally except that there is an increasing probability of a given cell being trapped in G2. In contrast, at high doses, over the first 24 h, a large proportion of cells are slowed down or blocked in S phase. Clearly, low doses of VP16-213 can modulate cell cycle distribution, enhancing topoisomerase II presentation, whereas at the higher doses the recruitment would require a considerable period (>24 h) to become optimal, reflecting the schedule dependency seen in the clinic.

Modulation of Topoisomerase II by Growth Factor Stimulation

It has been reported that estrogen potentiates the cytotoxicity of VP16-213 (16) in a breast tumor cell line and that potentiation does not depend upon cellular commitment to DNA synthesis (17). These findings raise the possibility that estrogen potentiates VP16-213 action by diminishing the size of a resistant G1 subpopulation. Flow cytometry can be used to examine the putative activation of such subpopulations with respect to DNA topoisomerase II availability.

Fig. 8a shows that cell samples exposed to FITC-conjugated antibody alone display only background levels of fluorescence. Cell nuclei rinsed in high salt buffer to remove matrix-bound topoisomerase II (Fig. 8b) similarly exhibit negligible antibody binding. Estrogen-deprived populations contain a small proportion of antibody-positive cells (Fig. 8c). Estrogen-treated samples exhibit a marked increase in the proportion of G1-phase antibody-positive cells (Fig. 8d) with the emergence of a discrete subpopulation. Such observations raise the possibility of practical clinical strategies (17) such as hormonal priming where sensitization would occur within a defined tissue type and arise from cellular activation rather than growth stimulation.

DISCUSSION

DNA topoisomerases are now recognized as important cellular targets for some cytotoxic drugs. Furthermore, high levels of the type II enzyme are found in proliferating tumor cells and the enzyme may be useful as a marker of proliferative intent. Using antibodies directed towards topoisomerase II, it is possible to map the availability of this target enzyme in a tumor cell population and infer the proliferation status and potential drug sensitivity of subpopulations. Thus, provision of target enzyme profiles for tumors could assist in the effective management of therapy. Drug-induced cell cycle delay, reflecting the antiproliferative action of a topoisomerase II poison, is dependent upon the availability of the target enzyme and the mode of trapping. Understanding the subtle interactions between the DNA helix, the topoisomerase II molecule and a given drug will clearly help to define the mechanism of enzyme trapping and identify factors which give rise to complex persistence and perhaps drug resistance. Indeed, many groups are now engaged in determining the cellular processes by which cancer cells can become either sensitive or resistant to topoisomerase inhibitors. In the

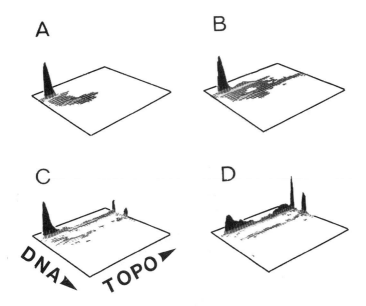

Figure 8 a-d. Flow cytometric analysis of topoisomerase II vs DNA content in breast tumor cells: a) Background fluorescence in nuclei stained with second antibody alone. b) Nuclei treated with 350 mM NaCl buffer to remove topoisomerase II prior to antibody incubation. c) Estrogen-deprived samples showing only a small proportion of immunoreactive cells. d) Estrogen-stimulated cells showing the emergence of a large GI subpopulation containing elevated levels of topoisomerase II. (Reprinted with permission from ref. 18).

future, cytometric techniques must be designed for indicating the subnuclear disposition of topoisomerase II with respect to regions of the genome undergoing specific metabolic processes, particularly DNA replication, sensitive to inhibition by anticancer drugs. The application of such "proximity" assays will not only facilitate our understanding of how topoisomerase inhibitors can act to kill tumor cells but also shed light on how the control of DNA conformation may facilitate function.

REFERENCES

1. Wang IC: DNA topoisomerases. Annu. Rev. Biochem. 54:665-697, 1985.
2. Liu LF: DNA topoisomerase poisons as antitumour drugs. Annu. Rev. Biochem. 58:351-375, 1989.
3. Smith PJ: DNA topoisomerase dysfunction: A new goal for antitumor chemotherapy. BioEssays 12: 167-172, 1990.
4. Kapusinski J, Darzynkiewicz Z, Traganos F, Melamed MR: Interaction of a new antitumour agent, 1-4-dihydroxy-5,8-bis[[[2-[2-hydroxyethyl]amino]ethyl]amino]-9,10-anthracenedione, with nucleic acids. Biochem. Pharmac. 30:231-240, 1981.
5. Long BH, Musial ST, Brattain MG: Comparison of cytotoxicity and DNA breakage activity of congeners of podophyllotoxin including VP16-213 and VM26: A quantitative structure-activity relationship. Biochem. 23:1183-1188, 1984.
6. Wozniak AJ, Ross WE: DNA damage as a basis for 4'-demethylepipodo-phyllotoxin-9-(4,6-O-ethylidene-2-d-glucopyranoside) (etoposide) cytotoxicity. Cancer Res. 43:120-124, 1983.
7. Lonn U, Lonn S, Nylen U, Winblad G: Altered formation of DNA in human cells treated with inhibitors of DNA topoisomerase II (Etoposide and Teniposide). Cancer Res. 49:6202-6207, 1989.
8. Glisson B, Gupta R, Hodges P, Ross W: Cross resistance to intercalating agents in an epipodophyllotoxin-resistant Chinese hamster ovary cell line: Evidence for a common intracellular target. Cancer Res. 46:1939-1942, 1986.
9. Minford J, Pommier Y, Filipski J, et al: Isolation of intercalator-dependent protein-linked DNA strand cleavage activity from cell nuclei and identification as topoisomerase II. Biochemistry 25:9-16, 1986.
10. Smith PJ, Makinson TA: Cellular consequences of overproduction of DNA topoisomerase II in an ataxia-telangiectasia cell line. Cancer Res 49:1118-1124, 1989.
11. Epstein RJ, Watson JV, Smith PJ: Subpopulation analysis of drug-induced cell-cycle delay in human tumour cells using 90° light scatter. Cytometry 9: 349-358, 1988.
12. Smith PJ, Anderson CO, Watson JV: Predominant role for DNA damage in etoposide-induced cytotoxicity and cell cycle perturbation in human SV40-transformed fibroblasts. Cancer Res. 46:5641-5645, 1986.
13. Fox ME, Smith PJ: Long-term inhibition of DNA synthesis and the persistence of trapped topoisomerase II complexes in determining the toxicity of the antitumour DNA intercalators mAMSA and mito-

xantrone. Cancer Res. 50:5813-5818, 1990.

14. Heck MM, Hittelman WN, Earnshaw WC: Differential expression of DNA topoisomerases I and II during the eukaryotic cell cycle. Proc. Natl. Acad. Sci. USA 85:1086-1090, 1988.

15. Slevin ML, Clark PI, Joel SP, et al: A randomised trial to evaluate the effect of schedule on the activity of etoposide in small cell lung cancer. J. Clin. Oncol. 7:1333-1340, 1989.

16. Epstein RJ, Smith PJ: Estrogen-induced potentiation of DNA damage and cytotoxicity in human breast cancer cells treated with topoisomerase II interactive antitumour drugs. Cancer Res. 48:297-303, 1988.

17. Epstein RJ, Smith PJ: Mitogen-induced topoisomerase II synthesis precedes DNA synthesis in human breast cancer cells. Biochem. Biophys. Res. Commun. 160:12-17, 1989.

18. Epstein RJ, Smith PJ, Watson JV, et al: Oestrogen potentiates topoisomerase-II-mediated cytotoxicity in an activated subpopulation of human breast cancer cells: Implications for cytotoxic drug resistant in solid tumours. Int. J. Cancer 44:501-505, 1989.

15

IN SITU HYBRIDIZATION OF CENTROMERIC DNA AS VISUALIZED
ON THE NUCLEAR MATRIX BY LASER CONFOCAL CYTOMETRY

Kenneth J. Pienta

INTRODUCTION

An average mammalian nucleus contains approximately
2 meters of DNA in its extended first order structure.
The packing of this length of DNA into a nucleus of only
10 μm diameter presents mammalian cells with a formidable topological packaging problem since the total length
of cellular DNA must be reduced about 10,000-fold to fit
within the confines of a single nucleus (1). Despite
this tremendous DNA packing ratio, DNA contained within
nuclei must have a dynamic conformation conducive to an
active role in a variety of biologic processes. For
example, the replication of the DNA occurs in 30,000 to
90,000 small units termed "replicons" that are synthesized in a precise order and temporal sequence. During
DNA synthesis, each of these DNA replicon units must be
copied by passing through a very large multienzyme
replicating complex (5×10^6 d) that contains the integrated biochemical site for DNA synthesis; these enzyme
complexes have been termed "replisomes" (2,3). Because
each of these individual replicon units of 50-100 Kbp
must pass through the fixed site of replication within a
30 min period, the double helix of DNA must be unwound
at a speed of over 100 rpm at each replicating site. On
average, DNA replication occurs at each site at 100
bp/min and there is a DNA turn in the helix for every 10
bp. Similarly, RNA polymerase II "reads" the DNA mole-

cule at a rate of 6,000 bp/min and can synthesize a mRNA
molecule of a 15 Kb gene in 3 min. The specific unwind-
ing of DNA units during replication and transcription
presents major topological considerations and a precise
ordering that is difficult to account for without evok-
ing a specific spatial organization and higher-order
structure within both nucleus and cell. To permit
diversity of function, DNA must be topographically and
topologically partitioned into independent functional
units or domains. This partitioning of DNA into domains
has been demonstrated in both neural cells and human-
hamster hybrid nuclei (4,5). This spatial and temporal
control of DNA is accomplished in part by the organization of DNA dictated by interactions with the nuclear
matrix. The nuclear matrix provides both the structural
support to the nucleus and the organizing sites for the
specific control of nucleic acid function as well as
directional, intranuclear, particulate transport (6,7,
Table 1).

The nuclear matrix is a framework scaffolding
forming the superstructure of the nucleus and consists
of peripheral lamins and pore complexes, an internal
ribonucleic protein network, and residual nucleoli (8-
10). The nuclear matrix is the protein framework on
which the DNA is organized into loop domains of both the
metaphase chromosome scaffold as well as the interphase
nuclear matrix and has been localized at the base of the
DNA loops where it exists in the wake of the DNA replicating fork (11,12).

In addition to the topological role of the nuclear
matrix, it has also been shown to be the site of many
critical cellular processes including the site for DNA
replication. The nuclear matrix has been demonstrated
to possess fixed sites of DNA synthesis which are located at the base of the loops and have been termed a
replitase. The replitase complex contains, at the

Table 1
Reported Functions Of The Nuclear Matrix

Nuclear Morphology
    The nuclear matrix contains structural elements of
the pore complexes, lamina, internal network, and nucleoli that contribute to the overall 3-dimensional organization and shape of the nucleus.

DNA Organization
    DNA loop domains are attached to nuclear matrix at
their bases and this organization is maintained during
both interphase and metaphase; nuclear matrix shares
some proteins with the chromosomal scaffold including
topoisomerase II, an enzyme that modulates DNA topology.

DNA Replication
    The nuclear matrix has fixed sites for DNA replica-
tion, containing the replisome complex for DNA replication that includes polymerase and newly replicated DNA.

RNA Synthesis and Transport
    Actively transcribed genes are associated with the
nuclear matrix; the nuclear matrix contains transcriptional complexes, newly synthesized heterogeneous nuclear
RNA, and small nuclear RNA; RNA-processing intermediates
(splicesomes) are bound to the nuclear matrix; mRNA is
transported on specific tracks within the nucleus involving nuclear matrix components.

Nuclear Regulation
    The nuclear matrix has specific sites for steroid
hormone receptor binding; the composition of the nuclear
matrix is cell and tissue specific. DNA viruses are
synthesized in association with the nuclear matrix; the
nuclear matrix is a cellular target for transformation
proteins and is altered by the transformation process.
At least one carcinogen has been shown to bind to the
nuclear matrix.

---

minimum, enzymes involved in DNA replication, including

DNA polymerase, topoisomerase, DNA methylase, dihydrofo-

late reductase, thymidylate synthetase, and ribonucleo-

side diphosphate reductase (3,13). Therefore, the

nuclear matrix plays both a structural and functional

organizing role in DNA regulation.

    The nuclear matrix also plays a central role in RNA

processing. Newly synthesized heteronuclear RNA and small nuclear RNA are both enriched on the nuclear matrix (14,15). The nuclear matrix has also been shown to be the site of attachment for products from RNA cleavage and for RNA processing intermediates; it has also been demonstrated to be the site of mRNA transcription (16-20). Active genes have been found to be associated with the nuclear matrix only in cell types in which they are expressed (21,22). Genes that are not expressed in these cell types are not found to be associated with the nuclear matrix. Further investigation into the association of active genes with the nuclear matrix has revealed a DNA loop anchorage site next to the enhancer region of several genes (23-28). Termed matrix associated regions (MARs) or scaffold attached regions (SARs), these sequences are usually approximately 200 base pairs in length, are A-T rich and contain topoisomerase cleavage sequences along with other sequences such as poly-adenylation signals. The MARs have also been shown to functionally confer transcriptional activity in genes in which they are inserted both in vivo and in vitro, and therefore, may be important carcinogen and/or mutagen targets (29,30).

Several observations indicate that the nuclear matrix is an important modulator of DNA organization and cell function. Nuclear matrix proteins vary in a cell type specific manner, suggesting that the nuclear matrix may play an important role in the tissue-specific, three-dimensional organization of DNA observed by in situ hybridization (31-35). The nuclear matrix interacts with steroid receptors to help modulate cell function (36,37). In cancer cells, transformation proteins appear to be associated with the nucleus, and many of these appear to be involved with the matrix. The nuclear matrix is one of the targets for retrovirus myc oncogene protein, adenovirus E1A-transforming protein, and polyoma large T antigen (38-43). Numatrin, a nuclear matrix protein, has been associated with the induction of mitogenesis and the nuclear matrix has been shown to be altered during cell transformation (33,44).

The nuclear matrix, therefore, is poised to play a critical and central role in cell structure and function. It structurally and functionally connects the DNA to the cell periphery. The importance of this structure in disease pathogenesis is poorly understood, however, techniques including in situ hybridization, are now available to start studying the nuclear matrix and its role in human disease.

MATERIALS AND METHODS

Nuclear Matrix Preparation (see Fig. 1)

This method is an adaptation of the original methodology of Vogelstein and colleagues (45) as well as Fey and Penman (34). Prostate adenocarcinoma DU-145 cells were plated on tissue culture slides (Nunc, Inc) and grown to near confluence. The cells were then treated with 0.5% Triton X-100 for five min to release the lipids and soluble proteins in a buffered solution containing 2mM vanadyl ribonucleoside, an RNAase inhibitor. Cells were then washed with phosphate-buffered saline (PBS). 0.25M ammonium sulfate with vanadyl ribonucleoside was then added to release the soluble cytoskeletal elements with salt for 10 min. DNAase-1 and RNAase-A at 25°C were then added to remove the soluble chromatin and RNA. The remaining in situ fraction contains the intermediate filaments and intact nuclear matrices.

In Situ Hybridization

Methods for in situ hybridization were modified from the techniques of Lawrence and colleagues (46-48). Hybridization of DNA on nuclear matrices is accomplished

232

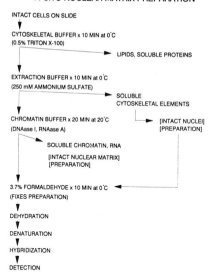

IN SITU NUCLEAR MATRIX PREPARATION

INTACT CELLS ON SLIDE

CYTOSKELETAL BUFFER x 10 MIN at 0˚C
(0.5% TRITON X-100)

→ LIPIDS, SOLUBLE PROTEINS

EXTRACTION BUFFER x 10 MIN at 0˚C
(250 mM AMMONIUM SULFATE)

→ SOLUBLE
CYTOSKELETAL ELEMENTS

CHROMATIN BUFFER x 20 MIN at 20˚C
(DNAase I, RNAase A)

→ [INTACT NUCLEI]
[PREPARATION]

SOLUBLE CHROMATIN, RNA
[INTACT NUCLEAR MATRIX]
[PREPARATION]

3.7% FORMALDEHYDE x 10 MIN at 0˚C ←
(FIXES PREPARATION)

DEHYDRATION

DENATURATION

HYBRIDIZATION

DETECTION

Figure 1. Schematic of the process for preparing nuclear matrices *in situ*. Intact cells are taken through a series of gentle extractions which remove the cell membrane as well as soluble proteins and DNA and RNA which are not tightly bound. The resulting *in situ* structure consists of the intermediate filament-nuclear matrix networks.

on the *in situ* prep as follows. Centromeric probes (biotinylated-11-dUTP) were obtained from Oncogene Science. For hybridization, samples are first denatured at 70°C for 2 min, then put into 70% followed by a 100% ethanol for 5 min each, then air dried. The hybridization solution is placed on samples and incubated at 37°C in a humidified chamber overnight. Hybridization is detected using avidin conjugated to fluorescein with residual nuclear matrix DNA being stained with propidium iodide.

Confocal Microscopy

Hybridized samples were visualized using an ACAS 570 confocal dual scanning laser microscope equipped with epifluorescent filters. This microscope is part of the Wayne State University School of Medicine Core Imaging Facility. Samples were visualized at a minimum of 1000x magnification using a 100x neofluor objective. Specimens were scanned mechanically at one micron intervals under microprocessor control along three axes through the confocal point to produce three-dimensional information. Multiple nuclei were examined and data were collected to ensure reproducibility.

RESULTS AND DISCUSSION

The advent of _in situ_ hybridization techniques has provided a new method for investigating nuclear matrix structure (49-53). The topological need for an internal organization of the nucleus would suggest that the nuclear matrix would be strongly associated with the centromeres of the chromosomes in the interphase nucleus. This would begin to ensure the three-dimensional localization of DNA within the nuclear volume. To test this hypothesis, _in situ_ nuclear matrix preparations of prostate cancer DU-145 cells were investigated for centromeric DNA-nuclear matrix interaction.

Fig. 2 demonstrates that centromeric DNA is tightly associated with the _in situ_ nuclear matrix. Utilizing the confocal laser cytometer, the different centromeric domains can be visualized throughout the interphase nucleus. Similar techniques can be used to visualize individual centromeres and chromosome domains (4,5,48, 54,55).

The nuclear matrix has been demonstrated to play an intimate role in the control of DNA organization and gene expression (33,56,57). The nuclear matrix is the structural element which determines, in part, the three-dimensional organization of the nucleus and is involved in DNA organization and function. _In situ_ hybridization can be used to learn more about this important nuclear structure, its role in cell function, and its role in

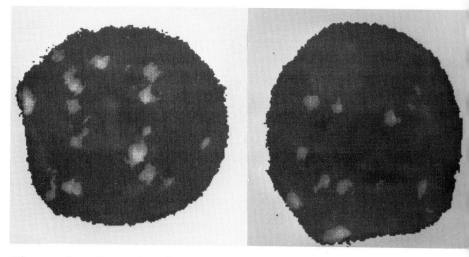

Figure 2. Scanning laser cytometer images taken 0.5 microns apart revealing centromeric DNA associated with the nuclear matrix of DU-145 human prostate cancer cells.

the organization of chromatin and chromosome structure.

ACKNOWLEDGEMENTS

The author would like to thank Mr. Jeff Lehr, Ms. Kerry Palmer, and Mr. Chuck Booth for technical help. The help of Ms. Mary Ann Terranova for assistance in arranging the manuscript is appreciated. This work was supported by NIH grant CA 22453 and the Wayne State University Fund for Medical Research and Education.

REFERENCES

1. Pienta KJ, Getzenberg RH, Coffey DS: Cell structure and DNA organization. CRC Rev. Eukaryotic Gene Expression 1:355-385, 1991.
2. Berezney R, Buchholtz LA: Dynamic association of replicating DNA fragments with the nuclear matrix of regenerating liver. Exp. Cell Res. 132:1-13, 1981.
3. Prem-Veer-Reddy G, Pardee AB: Inhibitor evidence for allosteric interaction in the replitase multi-enzyme complex. Nature 304:86-88, 1983.
4. Van Dekken H, Pinkel D, Mulliakn J, et al: Three-dimensional analysis of the organization of human chromosome domains in human and human-hamster hy-

brid interphase nuclei. J. Cell Sci. 94:299-306, 1989.

5. Manuelidis L, Borden J: Reproducible compartme-ntalization of individual chromosome domains in human CNS cells revealed by in situ hybridization and three-dimensional reconstruction. Chromosoma 96:397-410, 1988.

6. Nelson WG, Pienta KJ, Barrack ER, Coffey DS: The role of the nuclear matrix in the organization and function of DNA. Annu. Rev. Biophys. Biophys. Chem. 15:457-475, 1986.

7. Stuurman N, Meijne AML, van der Pol AJ, et al: The nuclear matrix from cells of different origin. J. Biol. Chem. 265:5460-5465, 1990.

8. Getzenberg RH, Pienta KJ, Coffey DS: The tissue matrix: Cell dynamics and hormone action. Endoc-rinol. Rev. 11:399-417, 1990.

9. Berezney R, Coffey DS: Identification of a nuclear protein matrix. Biochem. Biophys. Res. Commun. 60:1410-1417, 1974.

10. Pienta KJ, Coffey DS: A structural analysis of the role of the nuclear matrix and DNA loops in the organization of the nucleus and chromosome. J. Cell Sci. Suppl. 1:123-135, 1984.

11. Halligan BS, Small D, Vogelstein B, et al: Local-ization of type II DNA topoisomerase in nuclear ma-trix. J. Cell Biol. 99:128a, 1984.

12. Nelson WG, Liu LF, Coffey DS: Newly replicated DNA is associated with DNA topoisomerase II in cultured rat prostatic adenocarcinoma cells. Nature, 322:157-159, 1986.

13. Berezney R: The nuclear matrix: A heuristic model for investigating genomic organization and function in the cell nucleus. J. Cellular Biochem. 47:109-123, 1991.

14. Jackson DA, McCready SJ, Cook PR: RNA is synthe-sized at the nuclear cage. Nature, 292:552-555, 1981.

15. Harris SG, Smith MC: SnRNP core protein enrichment in the nuclear matrix. Biochem. Biophys. Res. Commun. 152:1383-1387, 1988.

16. Herman R, Weymouth L, Penman S: Heterogeneous nuclear RNA-protein fibers in chromatin depleted nuclei. J. Cell Biol. 78:663-679, 1978.

17. Miller SJ, Huang C, Pogo AO: Rat liver nuclear skeleton and ribonucleoprotein complexes containing hnRNA. J. Cell Biol. 76:675-691, 1978.

18. Mariman EC, Van Venrooij WJ: The nuclear matrix and RNA-processing use of human antibodies. In: Nuclear Envelope Structure and RNA Maturation. EG Smucker, GA Clawson (eds), New York, Alan R. Liss, Inc., pp.315-329, 1985.

19. Ciejek EM, Nordstromn JL, Tsai M, O'Malley BW: Ribonucleic acid precursors are associated with the

chick oviduct nuclear matrix. Biochemistry 21: 4945-4953, 1982.

20. Van Eekelen CAG, Van Venrooij WJ: hnRNA and its attachment to a nuclear protein matrix. J. Cell Biol. 88:554-563, 1981.

21. Robinson SI, Small D, Izerda R, et al: The association of transcriptionally active genes with the nuclear matrix of the chicken oviduct. Nucleic Acids Res. 11:5113-5130, 1983.

22. Robinson SI, Nelkin BD, Vogelstein B: The ovalbumin gene is associated with the nuclear matrix of chicken oviduct cells. Cell 28:99-106, 1982.

23. Cockerill PN: Nuclear matrix attachment occurs in several regions of the IgH locus. Nucleic Acids Res. 18:2643-2648, 1990.

24. Djondjurov L: DNA-RNA complexes that might represent transient attachment sites of nuclear DNA to the matrix. J. Cell Sci. 96:667-674, 1990.

25. Gasser SM, Laemmli UK: Cohabitation of scaffold binding regions with upstream/enhancer elements of three developmentally regulated genes of D. melanogaster. Cell 46:521-530, 1986.

26. Bode J, Maass K: Chromatin domain surrounding the human interferon-ß-gene as defined by scaffold-attached regions. Biochemistry 27:4706-4711, 1988.

27. Farache G, Razin SV, Wolny J, et al: Mapping of structural and transcription-related matrix attachment sites in the α-globin gene domain of avian erythroblasts and erythrocytes. Mol. Cell Biol. 10:5349-5358, 1990.

28. Stief A, Winter DM, Stratling WH, Sippel AE: A nuclear DNA attachment element mediates elevated and position-independent gene activity. Nature 341:343-345, 1989.

29. Gasser SM, Amati BB, Cardenas ME, Hofmann JFX: Studies on scaffold attachment sites and their relation to genome function. Int. Rev. Cytol. 119:57-96, 1989.

30. Ito T, Sakaki Y: Nuclear matrix association regions of rat α_2-macroglobulin gene. Biochem. Biophys. Res. Commun. 149:449-454, 1987.

31. Carter KC, Lawrence JB: DNA and RNA within the nucleus: How much sequence-specific spatial organization? J. Cell. Biochem. 47:124-129, 1991.

32. Getzenberg RH, Coffey DS: Tissue specificity of the hormonal response in sex accessory tissues is associated with nuclear matrix protein patterns. Molec. Endocrinol. 4:1336-1342, 1990.

33. Getzenberg RH, Pienta KJ, Huang EYW, Coffey DS: Identification of nuclear matrix proteins in the cancer and normal rat prostate. Cancer Res. 51: 6514-6520, 1991.

34. Fey EG, Penman S: Nuclear matrix proteins reflect cell type of origin in cultured human cells. Proc.

Natl. Acad. Sci. USA 85:121-125, 1988.

35. Fey EG, Wan KM, Penman SP: Epithelial cytoskeletal framework and nuclear matrix-intermediate filament scaffold: Three-dimensional organization and protein composition. J. Cell Biol. 98:1973-1984, 1984.

36. Barrack ER: The nuclear matrix of the prostate contains acceptor sites for androgen receptors. Endocrinol. 113:430-432, 1983.

37. Barrack ER, Coffey DS: Biological properties of the nuclear matrix: Steroid hormone binding. Recent Prog. Hormone Res. 38:133-195, 1982.

38. Buckler-White A, Humphery GW, Pigiet J: Association of polyoma T antigen and DNA with the nuclear matrix from lytically infected 376 cells. Cell 22:37-46, 1980.

39. Sarnow P, Mearing P, Anderson CW, et al: Identification and characterization of an immunologically conserved adenovirus early region 11,000 M_r protein and its association with the nuclear matrix. J. Mol. Biol. 162:565-583, 1982.

40. Staufenbiel M, Deppert W: Different structural systems of the nucleus are targets for SV40 large T antigen. Cell 33:173-181, 1983.

41. Moelling K, Benter T, Bunte T, et al: Properties of the myc-gene product: Nuclear association, inhibition of transcription and activation in stimulated lymphocytes. Curr. Top. Microbiol. Immunol. 113:198-207, 1984.

42. Eisenman RN, Tachibana CY, Abrams MD, Mann SR: V-myc and C-myc encoded proteins are associated with the nuclear matrix. Mol. Cell Biol. 5:114-126, 1985.

43. Verderame MF, Kohtz DS, Pollack RE: 94,000- and 100,000-molecular-weight simian virus 40 T-antigens are associated with the nuclear matrix in transformed and revertant mouse cells. J. Virol. 46: 575-583, 1983.

44. Feuerstein N, Chan PK, Mond JJ: Identification of numatrin, the nuclear matrix protein associated with induction of mitogenesis, as the nucleolar protein B23. J. Biol. Chem. 263:10608-10612, 1988.

45. Vogelstein B, Pardoll DM, Coffey DS: Supercoiled loops and eucaryotic DNA replication. Cell 22: 79-85, 1980.

46. Trask B, Pinkel D, Van Den Engh G: The proximity of DNA sequences in interphase nuclei is correlated to genomic distance and permits ordering of cosmids spanning 250 kilobase pairs. Genomics 5:710-717, 1989.

47. Lawrence JB, Taneja K, Singer RH: Temporal resolution and sequential expression of muscle-specific genes revealed by in situ hybridization. Developmental Biol. 133:235-247, 1989.

48. Lawrence JB, Singer RH, McNeil JA: Interphase and metaphase resolution of different distances within the human dystrophin gene. Science 249:928-932, 1990.

49. Nakayasu H, Berezney R: Mapping replication sites in the eukaryotic cell nucleus. J. Cell Biol. 108:1-11, 1989.

50. Berezney R: Visualizing DNA replication sites in the cell nucleus. Seminars in Cell Biol. 2:103-115, 1991.

51. Xing YG, Lawrence JB: Preservation of specific RNA distribution within the chromatin depleted nuclear substructure demonstrated by in situ hybridization coupled with biochemical fractionation. J. Cell Biol. 112:1055-1063, 1991.

52. Lawrence JB, Singer RH, Marselle LM: Highly localized tracks of specific transcripts within interphase nuclei visualized by in situ hybridization. Cell 57:493-502, 1989.

53. Pienta KJ: Techniques for the isolation of the nuclear matrix. Cytometry, in press.

54. McNeil JA, Johnson CV, Carter KC, et al: Localizing DNA and RNA within nuclei and chromosomes by fluorescent in situ hybridization. Genetic Anal. Tech. Applicat. 8:41-58, 1991.

55. Lichter P, Cremer T, Tang CJ, et al: Rapid detection of chromosome 21 aberrations by in situ hybridization. Proc. Natl. Acad. Sci. USA 85:9664-9668, 1988.

56. Pienta KJ, Partin AW, Coffey DS: Cancer as a disease of DNA organization and dynamic cell structure. Cancer Res. 49:2525-2532, 1989.

57. Getzenberg RH, Pienta KJ, Ward WS, Coffey DS: Nuclear structure and the three-dimensional organization of DNA. J. Cell Biochem. 47:289-299, 1991.

16

MEGAKARYOCYTOPOIESIS 2000

Alexander Nakeff

INTRODUCTION

One of the major lines of hematopoietic stem cell commitment is to the production of megakaryocytes and platelets. Its importance rests ultimately in the homeokinetic release of platelets into the peripheral blood. Circulating platelets are crucial to the maintenance of homeostasis and vascular integrity and play a role in tumor cell metastasis. Platelets are derived as anuclear fragments of cytoplasm from megakaryocytes resident mainly in bone marrow.

Both megakaryocytes and platelets present a formidable challenge to the application of multivariate analysis and fluorescence-activated cell sorting in structure-function studies designed to elucidate developmental pathways. The infrequency of megakaryocytes ($5x10^{-4}$ in marrow), their large size (20 to 50μ diameter), extreme fragility, low viability, high aggregation and extreme polyploidization significantly increase the complexity of rare-event analysis. The even more infrequent nature of their committed progenitor cells and the relative paucity of fluorescent probes that target unique (antigenic) structures in these progenitors renders their study extremely difficult. Platelets, on the other hand, are extremely small ($<1\mu$ diameter) and exquisitely sensitive to activation, aggregation and

degeneration; properties that demand strict control of
sampling and flow analysis techniques. A recent, comprehensive review of the application of multivariate
flow cytometry to megakaryocytopoiesis should be con-
sulted for details (1).

CYTOKINETICS

A scheme of murine stem cell-megakaryocyte-platelet
development is presented in Fig. 1. Diploid CFU-S that
become irreversibly committed to megakaryocyte differentiation, enter a series of interrelated clonogenic
progenitor cell compartments of decreasing proliferative
activity (i.e. large colony BFU-Meg with differentiation
to small colony CFU-Meg) under cytokine control (combined SCF + IL-3 + GM - CSF). Following this, a switch
from mitosis to NER leads first to the production of
morphologically-unidentified progenitors with ploidy
values possibly as high as 8C [that in the mouse can be
stained as SAChE cells], and then to early, polyploid
(8C) megakaryocytes capable of both subsequent NER to
increase polyploid level to 16C and 32C cells and cytoplasmic maturation leading to platelet formation.

PROGENITOR CELLS

The flow approach to these early megakaryocytic
developmental stages has utilized multivariate analysis
and sterile sorting of progenitors. Surface membrane
immunofluorescence (SMI) has utilized both monoclonal
and polyclonal specific antibodies to characterize
progenitors in unfractionated marrow as well as marrow
preparations enriched for these target cells. Electronic gating using combinations of SMI plus forward and
side (90°) light scatter, together with DNA, have been
used to sterile sort progenitors for subsequent clonogenicity testing and differentiation in vitro.

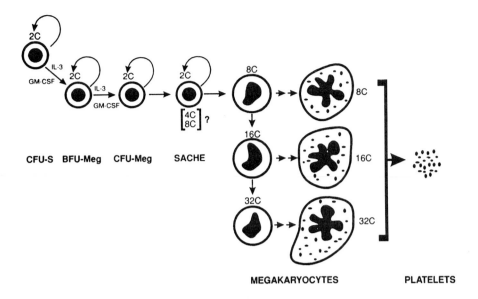

MEGAKARYOCYTES   **PLATELETS**

Figure 1. Model of stem cell commitment to megakaryo-cytopoiesis in murine bone marrow. A fraction of plu-ripotent stem cells (colony-forming unit, spleen or CFU-S) differentiate into clonogenic progenitors capable of extensive proliferation (burst-forming unit, megakaryo-cyte or BFU-Meg) and more limited proliferation (colony-forming unit, megakaryocyte or CFU-Meg). All of these progenitors are considered to be diploid 2C DNA content) and morphologically unidentified. With the commencement of nuclear endoreduplication (NER), a population of small acetylcholinesterase-positive (SAChE) cells (that may attain up to 8C polyploidy) differentiate into 8C megakaryocytes which can then undergo either one or more rounds of NER, leading to 16C and 32C megakaryocytes or cytoplasmic platelet formation.

As shown in Fig. 2, human progenitors have been analyzed primarily as non-adherent, low-density, T-cell depleted (NALT⁻) marrow sub-populations, using multiv-ariate flow analysis and multicolor SMI. BFU-Meg have been shown to be CD34$^+$ and HLA-DR$^-$ (2); whereas, CFU-Meg with a lower proliferative potential have been charac-terized as My10$^+$/HLA-DR$^+$ (2,3), RFB1$^+$/HLA-DR$^+$ (4), GpIIIa$^+$ (5), and GpIIb/IIIa$^+$ (6). Murine megakaryocytic progenitors have been less well-studied: They exhibit low Hoechst 33342 DNA staining (7) and bind fluorescei-

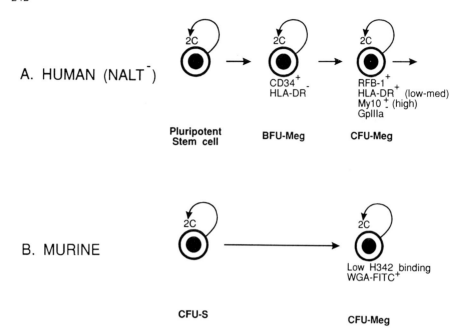

Figure 2. Surface membrane immunofluorescent (SMI) analysis of megakaryocyte progenitors in A) human and B) murine marrow, all assumed to be diploid (2C). Abbreviations: NALT, non-adherent, low-density, T cell-depleted; H342, Hoechst 33342 vital DNA stain; WGA-FITC, fluoresceinated wheat-germ agglutinin.

nated wheat-germ agglutinin (WGA-FITC) (8).

Although the above studies have provided useful markers for flow identification of rare progenitors, the markers are by no means specific. The lack of specificity seriously impedes the utility of flow cytometry to define early events in megakaryocytopoiesis. This problem not only reflects the low specificity of SMI, but also the paucity of progenitors which generally number about 1/10th the total number of megakaryocytes or about 1 progenitor per 20,000 total nucleated cells in marrow, a level that clearly qualifies as rare-event analysis.

A reasonable approach to the problem in the future involves more than two-color analysis, possibly going to three and four colors with multiple detectors and lasers. On the other hand, a more fundamental issue may be a "Catch-22" situation where, a decrease in the number and expression of commitment-specific maturation markers may reflect simply a decrease in differentiation from the morphologically-identified megakaryocyte to its earliest progenitor. However, pluripotent stem cells that are not yet committed to any lines of differentiation can be identified easily by their CD34[+] SMI positivity (6). Future directions to enhance the discrimination of multivariate analysis will involve:

a) The discovery of additional progenitor cell differentiation markers,

b) enhanced fluorescence background discrimination and low level detection of marker expression as well as the expression of low-affinity markers,

c) the judicious combination of functional markers such as acetylcholinesterase expression (for murine megakaryocytopoiesis) coupled with DNA and SMI probes,

d) the application of more discriminating pre-enrichment techniques to significantly improve the rate of detection and,

e) other strategies for enhancing rare event detection, as discussed in detail in this book (cf. chapter by J. Leary in this book on rare-event strategies).

MEGAKARYOCYTES

These cells present formidable problems to the optimal use of multivariate flow analysis, necessitating specific modifications to physical pre-enrichment techniques. These include unit-gravity sedimentation and counterflow centrifugal elutriation (CCE) for cell size separation, both continuous and discontinuous density gradient centrifugation using albumin, Ficoll and Percoll, ristocetin-coated magnetic beads for selective aggregation/disaggregation of megakaryocytes and combinations of the above. Over a period of about 10 years, these approaches have culminated in the enrichment of human and murine marrow megakaryocytes to near homogeneity with little or no selective cell loss as described in detail in several recent reviews (1,9).

We have applied discontinuous density gradient centrifugation followed by CCE in combination with multivariate flow cytometric analysis and cell sorting in a series of published studies (10-13). Fig. 3 presents a typical flow-generated DNA histogram on a partially-purified marrow preparation of rat megakaryocytes indicating the major polyploid peaks (8, 16 and 32C), as well as residual 2C and 4C cells. The contour plot of these cells (Fig. 4) as correlated data of DNA content

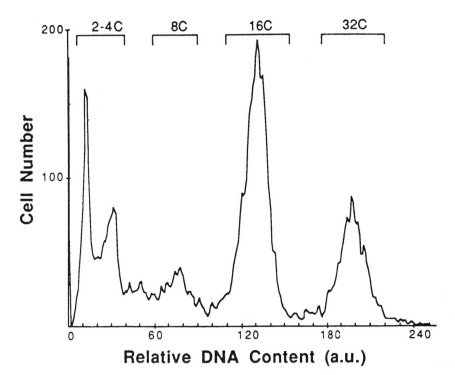

Figure 3. Representative DNA profile of magakaryocytes isolated by combined discontinuous Percoll density gradient centrifugation and CCE. Brackets above peaks indicate ploidy values. Hoechst 33342, DNA-related fluorescence is plotted on the abscissa [flow cytometer channel numbers 0-256, linear scale, arbitrary units (a.u.)] vs cell number plotted on the ordinate, linear scale.

versus cell size (FLS) indicates the heterogeneity of
cell size among the polyploid megakaryocytes which can
be used to sort various subsets for molecular analysis.
We demonstrated that although there was a linear relationship between (a) megakaryocyte DNA content (polyploidy) and cell size (FLS) and (b) DNA content and SMI,
using a specific polyclonal surface antigen, there
clearly was a non-linear relationship between FLS and

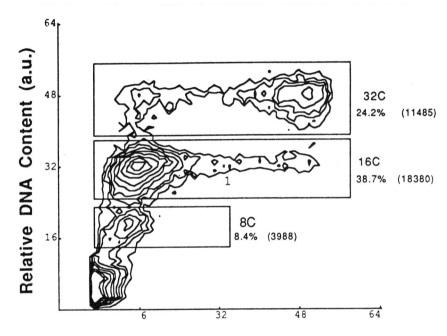

Forward Light Scatter (a.u.)

Figure 4. Contour plot of magakaryocytes in Fig. 3
illustrating the relationship between increasing DNA
content and FLS (<3 degrees, forward light scatter,
a.u.). Areas enclosed in boxes indicate ploidy classes
as marked. Number of cells at each ploidy level are
represented by topographic lines. Numbers in brackets
represent the absolute number of megakaryocytes within
each ploidy class which also are expressed as a per cent
of the total number of polyploid megakaryocytes. Numbers in brackets represent the absolute number of megakaryocytes in each ploidy class which also is expressed
as a percent of the total number of polyploid megakaryocytes.

SMI (10). These data supported the conclusion that although megakaryocytes increased their size as they progressed through successive rounds of polyploidization, the expression of platelet-associated proteins reflected a more complex relationship, one indicating that the prerequisite synthesis of membrane protein essential for cytoplasmic platelet formation may vary among ploidy classes.

We subsequently investigated the anomaly further by examining megakaryocyte surface structure (11) and Platelet-Factor 4 (PF-4) distribution (12,13) among sorted ploidy classes. Again, it was clear that although the three major polyploid classes (8C, 16C, and 32C) were involved in cytoplasmic platelet formation, they showed ploidy-specific distributions of cytoplasmic surface structures (11). Low ploidy (8C) megakaryocytes displayed the highest percentage of microvilli (40%) with fewer 16C (30%) and 32C (10%) cells displaying these structures. However, the microvilli in 8C cells were shorter and broader than those in 16C and 32C cells which were larger and more filamentous. In addition, the incidence of megakaryocytes with a nodular surface structure (i.e. one more mature than microvilli) increased as ploidy increased from 8C (10%) to 16C (20%) to 32C (30%). To complicate matters further, 32C cells displayed a unique surface structure referred to as "the pancake" representing a thin veil of cytoplasm surrounding a very large nucleus; the latter may represent a fully mature cell that had lost most of its cytoplasm in the process of platelet release. The heterogeneity of surface structures among ploidy classes may represent stages of cytoplasmic platelet formation and indicates that although high ploidy megakaryocytes are chronologically older (as a result of increased rounds of NER), the relationship between their ploidy and the corresponding levels of cytoplasmic maturity as reflected in surface structure is complex.

We next attempted to relate the heterogeneity of SMI and surface structure (previously described by scanning electron microscopy) to platelet-associated changes measured within the cytoplasm. We applied transmission electron microscopy and immunogold labeling of PF-4 to examine PF-4 sub-cellular localization and quantitate its distribution as a function of the cytoplasmic maturity of ploidy-sorted megakaryocytes (12, 13). We found that labeled PF-4 was present in the most immature cells as small vesicular structures (SVS) (12) that were transported from the Golgi zone to the cytoplasm for incorporation into alpha-granules (AG). The mean percentage of AG-related SVS was significantly related both to ploidy alone (when controlling for maturation) with 16C>32C>8C and to maturation alone (controlling for ploidy) where stage 1<2<3 (i.e. least to most mature) (13). When relating differences in ploidy to those in maturation, the only significance was for 8C vs 16C and 16C vs 32C. When the quantitation was carried out for the final disposition state of PF-4 (namely labeled AG), there was a significant interaction effect of ploidy and maturation; that is, more mature megakaryocytes contained more labeled AG within the 16C and 32C ploidy classes, but not within the 8C class (13). These data again support the conclusion that cytoplasmic maturation does not proceed synchronously within each of the three major ploidy classes, for reasons that remain to be determined.

FUTURE DIRECTIONS

Clearly, multivariate flow cytometry has made significant strides in elucidating megakaryocytic cell development. A number of areas hold promise for future research:

First, the discovery of new cytoplasmic and SMI

markers of early megakaryocytic development is important. In the murine system, the successful application
of fluorescent substrates for AChE activity will be
invaluable since it has been well characterized histo-
chemically as an early and specific functional probe of
megakaryocytic differentiation. The investigation of
oncoproteins such as p53 in modulating early polyploidization may be interesting since labeled monoclonals may
prove useful in quantitating oncoprotein expression as
an early marker of stem cell commitment to megakaryocytes.

Second, the ability to unequivocally identify and
vitally sort BFU-Meg, CFU-Meg and SAChE cells will
permit:

a) their further biochemical and physiologic characterization,

b) the study of cell-cell interactions with stromal
and lymphoid regulatory cells, and,

c) their successful application in shortening the
period of platelet regeneration in bone marrow trans-
plants as well as their ability _in vitro_ to sustain
short-term platelet production.

Third, regulatory mechanisms remain to be defined
which are important in the progenitor transition from
mitosis to NER. Studies investigating the role of
cyclins in NER could utilize labeled monoclonal probes
for Ki-67 _vs_ PCNA to quantitate expression as a function
of cell cycle. The ability of the family of cytochala-
sins to act as specific mitotic inhibitors may be utilized to determine at which point in late G2 a partial
arrest may trigger NER _vs_ mitosis. The role of specific
cytokines that are proliferative (IL-3, GM-CSF), matu-
rational (IL-6, IL-11) and inhibitory (TGF-ß) could be
explored in single progenitor cell cultures. The role
of megakaryocytes to self-destruct in the process of
platelet formation with concomitant high levels of

polyploidy makes it attractive as an apoptotic cell model that could be probed by new flow cytometric approaches.

Fourth, an important area for future studies involves NER regulation in morphologically-identified megakaryocytes. The precise mechanism of platelet feedback on NER remains to be defined. The role of thrombopoietin as a specific, circulating humoral stimulator of NER will be examined following its isolation in recombinant form. The mechanism by which platelets inhibit stem cell commitment to megakaryocytopoiesis and suppress NER may best be examined by multivariate flow analysis of megakaryocytes that are vitally stained for DNA content.

Fifth, the production of functional platelets may be studied in sophisticated marrow-stromal, long-term cell cultures. New techniques for detecting "hemostatically-active" platelets by flow cytometry will provide some answers as to whether this population of newly-formed platelets is the important target for triggering both positive and negative regulators. In addition, we will learn how important the derivation of platelets from a particular ploidy class of megakaryocyte is to the former cells' subsequent physiologic function.

In conclusion, the future of flow and image cytometry in enhancing our knowledge of megakaryocyte development and its regulation appears promising.

ACKNOWLEDGEMENTS

I wish to thank D. Liang and M. KuKuruga for their technical support and Drs. E. Hegyi and M. Weller for their contributions during post-Doctoral and WSU Doctoral programs, respectively. Partial financial support was provided by the WSU Ben Kasle Trust for Cancer Research.

REFERENCES

1. Nakeff, A: Megakaryocytopoiesis: Analysis and Cell Sorting. In: Flow Cytometry in Hematology, OD Laerum and R Bjerknes (eds), Academic Press, New York, NY, pp. 43-58, 1992.

2. Lu L, Walker D, Broxmeyer HE, et al: Characterization of adult human marrow hematopoietic progenitors highly enriched by two color cell sorting with MY10 and major histo-compatibility class II monoclonal antibodies. J. Immunol. 139:1823-1829, 1987.

3. Watt SM, Katz FE, Davis L, et al: Expression of HPCA-1 and HLA-DR antigens on growth factor and stroma-dependent colony forming cells. Brit. J. Hematol. 66:153-159, 1987.

4. Bodger MP, Izaquirre CA, Blacklock HA, Hoffbrand AV: Surface antigenic determinants on human pluripotent and unipotent hematopoietic progenitor cells. Blood 61:1006-1010, 1983.

5. Kanz L, Mielke R, Fauser A: Analysis of human hemopoietic progenitor cells for the expression of glycoprotein IIIa. Exp. Hematol. 16:741-747, 1988.

6. Debili N, Issaad C, Masse JM, et al: Expression of CD34 and platelet glycoprotein during human megakaryocytic differentiation. Blood 80:3022-3035, 1992.

7. Pallavicini MG, Levin J, Summers L, Levin F: Multivariate flow cytometric characterization and enrichment of murine megakaryocyte progenitors, CFU-Meg. Exp. Hematol. 15:704-709, 1987.

8. Bauman JGL, Chen MG: Light scatter and cell surface progenitors of murine megakaryocyte progenitor cells. Exp. Hematol. 15:1074-1079, 1987.

9. Nakeff A: Megakaryocytic Cells. Bibl. Hematol. 48:131-209, 1984.

10. Worthington RE, Nakeff A, Micko S: Flow cytometric analysis of megakaryocyte differentiation. Cytometry 5:501-508, 1984.

11. Weller MA, Szela J, Barnhart MI, Nakeff A: Surface ultrastructure of human megakaryocytes sorted on the basis of DNA content. Scan. Electron Microscopy 4:1437-1443, 1986.

12. Hegyi E, Nakeff A: Ultrastructural localization of platelet factor 4 in rat megakaryocytes by gold-labeled antibody detection. Exp. Hematol. 17:223-228, 1989.

13. Hegyi A, Heilbrun LK, Nakeff A: Immunogold probing of platelet factor 4 in different ploidy classes of rat megakaryocytes sorted by flow cytometry. Exp. Hematol. 18:789-793, 1990.

17

THE MOLECULAR BASIS FOR THE CONTROL OF MAMMALIAN CELL
GROWTH

Gerard I. Evan

INTRODUCTION

Multicellular organisms face a unique problem in
regulating proliferation of their component cells. This
is because, whereas failure of even a large proportion
of cells to divide within a tissue will generally be of
little consequence as there are always other cells
present that can renew the affected tissue, the unrestrained proliferation of even one cell and its progeny
will be lethal. Indeed, this is the disease we call
cancer. Recently, there have been many spectacular
advances in our understanding of the molecular processes
that regulate cell proliferation. This review will
outline our current understanding of the major features
of cell growth regulation in mammalian cells.

AN HEIRARCHY OF CONTROL

Mammalian cell proliferation comprises three linked
processes:

1. The cell cycle, whereby DNA synthesis and cell
division alternate and it is ensured that one process
finished before the other starts. As all cells have to
have a cell cycle, it is self evident that the machinery
regulating the cell cycle is likely to be substantially
conserved throughout evolution, irrespective of unicellular or a multicellular nature. Accordingly, many
recent advances in our understanding of the cell cycle

have come from studies on unicellular yeast: The budding yeast Saccharomyces <u>cerevisiae</u> and the fission yeast Schizosaccharomyces <u>pombe</u>.

2. The regulation of proliferation. This is arguably the key growth control step in normal mammalian cells, is presumably damaged in tumour cells and is likely to be absent or extensively modified and adapted in unicellular eukaryotes. Normal untransformed mammalian cells are the only effective model in which to study this process.

3. Control of tissue mass. The homeostatic control of tissue mass and tissue architecture relies on the net gain or loss of cells from tissues. This is a dynamic process that includes those processes leading to cell gain (i.e. cell cycle and proliferation control), cell stasis (via quiescence and differentiation) and cell loss (shedding, cell death).

In seeking to understand how these three processes might be interlinked at the molecular level, it is first necessary to examine what is known of the molecular machinery mediating each one. I will therefore begin by discussing the cell cycle, control of proliferation and regulation of tissue mass in turn and try to show how each appears to be integrated with the others.

The Cell Cycle

The cell cycle is, as already outlined, necessarily common to all cells and, for this reason, much of the molecular machinery that regulates it is also highly conserved through evolution. Much of our current understanding of the regulation of the cell cycle comes from the study of temperature-sensitive (ts) cell division cycle (called "<u>cdc</u>" mutants in <u>pombe</u> and "CDC" in <u>cerevisiae</u>) mutants in yeast which are unable to proceed through the cell cycle above a permissive temperature (1). The first of these to be characterised in S. <u>pombe</u>

was cdc2 (equivalent to CDC28 in cerevisiae). cdc2 mutants arrest at both the G1/S and G2M boundaries implying that regulation of both of these cell cycle checkpoints requires a common gene product. cdc2 encodes a 34 kDa serine/threonine protein kinase, p34^{cdc2}, that is substantially conserved throughout eukaryotic evolution and that phosphorylates multiple intracellular target proteins. Given its key role in regulating both the G1/S and G2/M transitions, it is at first site paradoxical that levels of p34^{cdc2} are invariant through-out the cell cycle. Later studies, however, showed that p34^{cdc2} kinase activity is detectable only at the G1/S and G2/M boundaries. This is because p34^{cdc2} is active only when associated with members of a family of protein called cyclins. Cyclins, as their name implies, were initially identified as proteins demonstrating a cell cycle-dependent pattern of expression. Thus, p34^{cdc2} activity is cell-cycle-dependent because it is regulated by a cell cycle-dependent cyclin subunit. What, then, regulates the cyclins? The answer to this is best illustrated by examining the cell cycle transition that is currently best understood, G2/M.

Control of the G2/M Transition

The cyclins mediating the G2/M transition are cyclins A and B. Several A- and B-type cyclins are known to exist but it is unclear whether each has a different specificity and function. Both A and B cyclins accumulate throughout S and G2 phases of the cell cycle, with accumulation of A slightly preceeding B (2). As each cyclin is synthesised, it complexes with available p34^{cdc2}. However, the complex is immediately inactivated via phosphorylation of p34^{cdc2} at position Tyr15 by the mitotic inhibitory kinase p1071^{wee1} and p66^{mik1}. p34^{cdc2}-cyclin A/B activity is thus held in abeyance until the end of G2, at which point p34^{cdc2} is dephosphorylated, and hence activated, via the $p180^{cdc25}$ phosphatase. Active $p34^{cdc2}$-cyclin A/B then initiates mitosis following which cyclins A and B are rapidly degraded. Indeed, evidence suggests that cyclin A and B degradation is <u>necessary</u> for complete exit from M (mitosis). What, then, activates $p180^{cdc25}$? The answer is presently unclear but some evidence suggests that $p180^{cdc25}$, itself, may be regulated by cyclin B.

Control of the G1/S Transition

Can we learn anything about the G1/S transition from what we know of regulation of G2/M? The answer appears to be yes. We now know that there are specific G1/S cyclins. In S. <u>cerevisiae</u> these have been identified as CLN 1, 2 and 3 and they share some 25% homology with cyclins A and B. CLN1 and 2 are inherently unstable and their levels are thus regulated principally by rates of transcription. The CLN1 and 2 genes possess a specific cell cycle control sequence (the Swi4/6-dependent cell-cycle box or SCB) which, together with another enhancer sequence (the MluI cell-cycle box or MCB) is also present in a number of other genes whose expression is regulated within G1 and S phase (see below). CLN3, on the other hand, is a stable protein. In yeast, expression of CLN1 and 2 is repressed by mating factor and this is thought to be responsible for the growth arrest induced by mating factor. Identification of the analogous G1 cyclins in S. <u>pombe</u> has proven difficult. To date, two candidates, the products of the two genes <u>puc</u>1+ and <u>cig</u>1+, are known.

In mammalian cells, commitment to proceed through the cell cycle maps to a late G1 "restriction point" (often called "R"), typically about two hours before the onset of DNA synthesis. R is commonly thought to represent the mammalian G1/S regulatory transition step, analogous to that at G2/M, and is presumably regulated by cyclins. Indeed, three mammalian G1/S cyclins are known, C, D and E (2,3). C peaks early in G1, E just before S. Both are expressed in a cell cycle-dependent manner. In contrast, expression of the D cyclins (D1, 2 and 3) appears to be dependent upon mitogen availability rather than cell cycle position, at least in macrophages in which they were first characterised (4). D cyclins may therefore represent p34^{cdc2}-controlling subunits whose expression is linked not to cell cycle position but to the decision-making machinery that governs whether or not a cell will enter cycle in the first place. D cyclins may thus constitute the bridge between control of the cell cycle and control of cell proliferation.

Regulation of Gene Expression during the Cell Cycle

Clearly, a defined pattern of gene expression is required during cell cycle progression. For example, the onset of DNA synthesis requires expression of proteins needed for DNA replication such as thymidylate kinase, thymidylate synthase, DNA polymerase and dihydrofolate reductase: Histones and some other chromatin components must be synthesised during S phase. In S. cerevisiae, genes for many of these G1/S-specific proteins contain the SCB or MCB boxes (described above) in their regulatory regions. The SCB box (5'-CACGAAA-3') is bound by the product of the SWI4 gene which, in association with the SWI6 product, causes activation of relevant genes (5). SWI6 associates with a different protein to bind the MCB box (5'-ACGCGT-3') and induces analogous activation. The S. pombe counterpart of SWI6 is the product of the cdc10 gene. In mammalian cells, the analogous transcription factor controlling G1/S genes is called E2F. E2F is itself the centre for much interest because it is found associated with cyclin A as cells proceed from G1 to S in a complex that also includes a 107 kDa nuclear protein (6). This p107 has homology with the product of the retinoblastoma gene, p105^{RB}, with which E2F is associated in G1 (7-9). Current dogma asserts that E2F is inactive when complexed, and it is the displacement of p105^{RB} by p107 that leads to E2F activation and thence to activation of appropriate genes. This overly simplistic model is certain to be further modified over the next few months or years, but it provides a useful basis for modelling cell cycle control of gene expression.

Anti-Proliferative Factors - Factors Restraining Cell Cycle Progression

Unrestrained cell proliferation is the ultimate danger in multicellular organisms. Not surprisingly, therefore, it appears that cell cycle progression is modulated not only by positive factors, that drive cell proliferation, but also by factors that restrain it. At least two such endogenous growth suppressors are known. One, the product of the retinoblastoma locus p105^{RB}, has already been mentioned above. p105^{RB} undergoes a cyclic phosphorylation and dephosphorylation during the cell cycle. At about the time of cell cycle commitment, p105^{RB} is phosphorylated, a process that is presumed to inactivate it (10,11). The function of active p105^{RB} is thus presumed to be to prevent entry into S phase. p105^{RB} is involved in formation of complexes with the E2F transcription factor as well as with a number of other transcription factors such as the nuclear products of the oncogenes c-myc (12), c-fos and c-jun. The other endogenous growth suppressor, p53, was first identified as a protein that complexed with the principal oncoprotein of the SV40 DNA tumour virus, large T antigen (13). p53 may also be involved in controlling entry into S phase although the mechanism is unclear.

CELL CYCLE AND CANCER

Perturbation of cell cycle components is associated
with neoplasia in a variety of ways. The cyclin A gene
is the frequent site of integration of the hepatitis B
virus, associated with hepatocellular carcinoma. Viral
insertion typically induces deregulated expression of
nearby genes and presumably deregulated cyclin A expression contributes to tumour formation in this instance
(14). Both the bcl-1 gene, translocated to the immunoglobulin gene heavy chain locus in many B cell lymphomas, and the Prad1 locus, translocated to the parathyroid hormone gene in parathyroid tumours, encode D type
cyclins. Translocations of this kind also cause deregulated expression of attendant genes, suggesting that
damage to control of cyclin D expression is also carci-
nogenic (14).

Both the retinoblastoma protein p105^{RB} and p53 are
intimately involved in human cancers. The retinoblastoma gene RB is functionally inactivated in human retinoblastoma as well as in a variety of other tumours (15)
suggesting that its growth-limiting properties act as an
effective bulwark against spontaneous carcinogenesis.
p105^{RB} is also involved in complexes with the large T
antigen of SV40, the E1A protein of adenovirus and the
E7 protein of human papilloma virus, all tumour viruses.
Current thinking is that p105^{RB} is functionally inactivated in such complexes and that, by promoting host cell
growth, facilitates attendant viral proliferation. p53,
which is functionally inactivated in some 60% of all
human tumours, also forms complexes with the tumour
antigens of DNA tumour viruses; specifically SV40 large
T, adenovirus E1B and human papilloma virus E6 (13).
Once again, sequestration and concomitant inactivation
of p53 is thought to confer an advantage on the virus
because by promoting host growth the virus promotes its
own.

Thus, controlled expression and activation of cell cycle regulatory components is an essential prerequisite for effective control of cell proliferation. With the possible exception of the D cyclins, the factors regulating the cell cycle are not in the main involved in determining whether or not a cell is in cycle or out of cycle (quiescent). Yet, the transition from quiescence to proliferation is, as discussed, perhaps the major control transition in mammalian cells. We will therefore next examine what is known of the molecular events that accompany exit from G0 and entry into cycle - the regulation of proliferation.

Regulation of Cell Proliferation

Notwithstanding the obvious importance in cellular growth control of positive and negative factors mediating transition through cell cycle checkpoints, many, perhaps most cells in an adult human appear not to be in cycle at all. Instead, they are quiescent. In many cases, quiescence arises as a result of terminal differentiation and irreversible loss of proliferative capacity. In other instances, however, quiescent cells retain full proliferative capacity and respond rapidly to appropriate mitogenic stimulation by entry into the cell cycle and eventual division. Such is the case in, for example, fibroblasts, lymphocytes and hepatocytes. Quiescent cells are not usually arrested at either of the cell cycle checkpoints but reside in a non-cycling state often termed G0. The transition from G0 into G1, therefore, constitutes a major point at which mammalian cell proliferation is regulated.

Immediate Early Genes

When a quiescent fibroblast or lymphocyte is stimulated with mitogens _in vitro_ there is an abrupt induction of some 40-100 genes (16-18), the so-called "imme-

diate early growth response genes." These genes encode
proteins many of which are presumably involved in en-
abling the cell to proliferate. Some are cytokines,
some elements of signal transduction pathways, some
structural and cytoskeletal proteins, and others are
transcription factors. Substantial attention has fo-
cussed on this latter group, the transcription factors,
because they are presumed to mediate genetic changes
required for proliferation. Amongst them are the products of the c-fos, c-jun and c-myc genes, all known to
have oncogenic properties if deregulated and all sequence-specific DNA-binding proteins. The overt proto-
oncogenic properties of the c-fos, c-jun and c-myc genes
suggests that they have crucial roles in regulating cell
proliferation and the decision whether to proliferate or
arrest.

Fos and Jun

Both the c-fos, c-jun genes were first identified
as viral oncogenes in the FBJ and FBR murine osteosarcoma viruses and in the ASV17 avian sarcoma virus, respectively. Both c-fos and c-jun are members of gene families that encodes a group of transcription factors which
includes c-jun, jun-B, jun-D, c-fos, fos-B and fra-1
(19,20). The various fos and jun proteins function as a
variety of homo- and hetero-dimers which form through a
characteristic dimerisation motif called a leucine
zipper which they all possess (21). The Fos-Jun dimers
constitute the AP-1 transcription factor activity originally isolated from nuclear extracts by its specificity
for the common TGACTCA motif and which confers inducibi-
lity by the phorbol ester TPA (20). The TGACTCA is
found upstream of a variety of genes activated during
the G0/G1 transition.

Although their oncogenic properties strongly imply
that both Fos and Jun proteins are intimately involved

in cell proliferation, it is clear that they can function in other contexts also. For example, both are
expressed in non-mitotic CNS neurones in response to
synaptic stimulation (22-25). Thus, their function
appears to be to act as generalised switches that can
initiate a variety of cellular responses: Precisely what
they do, however, depends upon the cellular context in
which they are expressed.

Myc

The c-myc proto-oncogene was originally isolated by
its homology to the transforming genes of the avian
myelocytic leukaemia viruses. C-myc is one member of a
family of myc genes in man which includes N-myc, L-myc,
B-myc, and S-myc (26), although only c-myc appears to be
expressed in adult cells. In untransformed cells, c-myc
expression is linked with growth state. It is absent in
quiescent cells but rapidly induced by mitogenic growth
factors (27). For this reason, it has frequently been
grouped with other immediate early growth response genes
such as c-fos and c-jun. However, whereas most other
immediate early genes are expressed transiently and only
at the G0/G1 interface (16) or in response to other
adventitious stimuli, c-myc appears to be expressed at a
constant rate throughout the cell cycle in proliferating
cells (28-31) but in a manner that is mitogen dependent
(31,32). As both the c-myc mRNA and protein have extremely short half-lives, continuous expression of c-myc
can only occur as a result of continuous synthesis.
This argues that c-myc exerts a continuous function
throughout cell proliferation.

Almost all tumour cells express steady state levels
of c-myc mRNA and protein which are substantially higher
than those seen in untransformed proliferating fibroblasts (33). Although the significance of this elevated
expression is unknown, altered regulation of the c-myc

gene, in particular by amplification and chromosomal translocation, has been implicated in the genesis of a number of human and non-human tumours (34). Interestingly, in many tumour cells that express high levels of c-myc protein there is no obvious genetic lesion to account for the over-expression. However, because c-myc expression appears to be autoregulating (35), over-expression of c-myc in tumour cells is unlikely to be simply a concomitant effect of some other oncogenic lesion. Growth factor deprivation does not, in general, lead either to growth arrest of tumour cells or to repression of c-myc expression as is the case in untransformed fibroblasts (36). C-myc expression alone will drive quiescent cells into cycle in the absence of mitogens (37), and keep them there (38,39, manuscript submitted). Moreover, specific anti-sense inhibition of c-myc expression effectively blocks entry of cells into cycle in response to mitogens (40). Myc is therefore both necessary and sufficient for cell proliferation. Clearly, this is a potentially very dangerous state of affairs because any mutation that deregulates c-myc expression will in principle be oncogenic. Recently, this paradox has been resolved by the discovery that Myc is a potent inducer of programmed cell death (38) - see below.

The c-myc gene product, Myc, is a sequence-specific DNA binding protein (41) that forms an active complex with a heterologous partner called Max (39,42). Dimerisation is mediated by a C-terminal domain, part of which shares structural homology with the leucine zipper present in Fos and Jun proteins described above. In addition, however, both Myc and Max possess another mutually interactive domain, a Helix-Loop-Helix (HLH) motif, that further stabilises dimerisation. Structures for the Myc-Max interaction have been modelled (43). The Myc-Max dimer is almost certainly a transcription

factor that modulates expression of specific genes that
then mediate the observed effects of Myc expression -
mitogenesis (37,38), transformation (44,45), autosuppression (35,46), inhibition of differentiation (47) and
apoptosis (38).

ROLE OF GROWTH RESPONSE GENES

From this brief survey of the characteristics of
the c-fos, c-jun and c-myc genes and their proteins, it
is clear that they fall into two categories. Fos and
Jun exemplify a class of transcription factor that
mediates a transition from one state (e.g. quiescence)
to another (e.g. proliferation) but is not required to
maintain the new state. Such factors are also involved
in a variety of other molecular switching events which
are not associated with cell proliferation. I suggest
that Fos and Jun are examples of generic "toggle switches" that flick a cell from one stable genetic state into
another. In contrast, c-myc exemplifies a switch that
must be continuously engaged in order to have its effect. I liken this to an "accelerator" which must be
continuously activated in order to maintain a response.
However, both Fos/Jun and Myc classes operate through
their own target genes and it is only by determining the
nature and function of these target genes that we will
uncover the biological role of these molecular switches.

Control of Tissues - The Problem of Integration

Tissues are dynamic structures whose architecture
and size depends upon the net behaviour of large numbers
of cells working in concert. How is such integration
achieved? In principle, there are four possible fates
for any one cell: Quiescence, proliferation, differentiation and death. I have discussed what is known about
quiescence and proliferation, but what of differentiation and death? In many ways, differentiation can be

seen as the genetic obverse of proliferation in that as
cells differentiate they almost always lose their capacity for self-renewal. Thus, growth and differentiation
are necessarily and inversely linked. In this scheme,
the onset of differentiation, mediated by interactions
with specific cytokines, automatically leads to the
cessation of proliferation. A common fate of the differentiated cell is eventual death. This can occur by a
variety of mechanisms, for example physical loss (e.g.
shedding of skin cells or loss of colonic epithelium
into the gut lumen), by accident (e.g. random poisoning,
lethal mutation, etc.) and also by programmed cell
death. One of the mechanisms of programmed cell death
is the process of apoptosis - an energy-dependent active
process involving a specific pathway of chromatin con-
densation and fragmentation, cellular vesicularisation
and activation of surface markers signalling phagocyto-
sis (48-53). Apoptosis is involved in the shaping of
tissues in birth embryo and adult and is also one of the
pathways of tumour cell destruction in response to
chemo- and radio-therapy (54,55).

Apoptosis

The genetics of apoptosis is best understood in the
nematode Caenorhabditis elegans in which it has been
shown to depend on the interplay of specific cell death
(ced) genes - in particular ced 3 and ced 4, which
induce death, and ced 9 which blocks it (56). In mammalian cells, the pathway is far less well defined. The
gene bcl-2, first identified as the site of reciprocal
translocations in human follicular lymphomas, is able to
block apoptosis (57-60) and may be a major regulator of
the pathway in mammals. Recently, the c-myc gene has
been shown to be a potent inducer of apoptosis (38)
whose lethal effects are inhibited by BCL-2 expression
(Fanidi, Harrington and Evan, in press). The c-myc gene

thus appears to sit in a pivotal position as a regulator not only of proliferation, but also of cell death. What might be the sense in having one intracellular component regulating two such contradictory processes? The answer seems to be that c-myc is, as outlined above, both necessary and sufficient for cell proliferation. Thus, any mutation deregulating c-myc expression would be oncogenic and lethal. If, however, in addition to regulating genes mediating proliferation, Myc also activates genes mediating apoptosis, then all cells expressing Myc would necessarily be primed for pro-grammed cell death. Successful proliferation would then presumably occur only if apoptosis were actively inhib-ited, perhaps by activation of complementary signal transduction pathways. The dual properties of Myc therefore serve to integrate the different tiers of control of cell proliferation and thereby give rise to a multicellular organism that can maintain its tissue architecture yet retain proliferative capacity in its cell necessary for repair and growth.

CELL GROWTH CONTROL - INTEGRATION OF A HEIRARCHY OF REGULATION

The control of mammalian cell growth thus comprises three interlocking tiers of control - cell cycle; pro-liferation; and, at the level of net cell growth in tissues. Although precise details of how this heirarchy is interconnected remain to be determined, it is begin-ning to be possible to construct a generalised model, subject, of course, to extensive revision in the future. In this model, control of the cell cycle is interlocked with control of proliferation by the positive effects of specific cyclins, probably the D cyclins, and by the negative anti-proliferative effects of factors such as the retinoblastoma and p53 proteins. The realisation that almost all human cancers involve loss of specific genes (61) suggests that many more such growth-limiting cellular factors await discovery. The control of proliferation is in the main governed by the components of the signal transduction pathways - this is the realm of the oncogenes and their products. Some of these oncogenes, however, (e.g. c-myc) appear to have dual functions, and simultaneously mediate the contradictory processes of cell proliferation and cell death. Net cell proliferation thus only occurs if the cells concerned receive yet further signals that mitigate the lethal effects of such ambivalent oncogenes and allow survival. The behaviour of individual cells is thus necessarily connected with the behaviour of a tissue as a whole, so providing the framework for large scale cellular integration that is the hallmark of multicellular organisms.

REFERENCES

1.   Bartlett R, Nurse P: Yeast as a model system for understanding the control of DNA replication in eukaryotes. Bioessays 12:457-463, 1990.
2.   Pines J: Cell proliferation and control. Curr. Opin. Cell Biol. 4:144-147, 1992.
3.   Pines J: Cyclins: Wheels within wheels. Cell Growth Diff. 2:305-310, 1991.
4.   Matsushime H, Roussel MF, Ashmun RA, Sherr CJ: Human D-type cyclin. Cell 65:701-713, 1991.
5.   Andrews B: Dialogue with the cell cycle. Nature 355:393-394, 1992.
6.   Mudryj M, Devoto SH, Hiebert SW, et al: Cell cycle regulation of the E2F transcription factor involves an interaction with cyclin A. Cell 65:1243-1253, 1991.
7.   Bagchi S, Weinmann R, Raychaudhuri P: The retinoblastoma protein copurifies with E2F-I, an E1A-regulated inhibitor of the transcription factor E2F. Cell 65:1063-1072, 1991.
8.   Chellappan SP, Hiebert S, Mudryj M, et al: The E2F transcription factor is a cellular target for the RB protein. Cell 65:1053-1061, 1991.
9.   Chittenden T, Livingston DM, Kaelin WJ: The T/E1A-binding domain of the retinoblastoma product can interact selectively with a sequence-specific DNA-binding protein. Cell 65:1073-1082, 1991.

10. Buchkovich K, Duffy LA, Harlow E: The retinoblastoma protein is phosphorylated during specific phases of the cell cycle. Cell 58:1097-1105, 1989.

11. Chen P-L, Scully P, Shew J-Y, et al: Phosphorylation of the retinoblastoma gene product is modulated during the cell cycle and cellular differentiation. Cell 58:1193-1198, 1989.

12. Rustgi AK, Dyson N, Bernards R: Amino-terminal domains of c-myc and N-myc proteins mediate binding to the retinoblastoma gene product. Nature 352: 541-544, 1991.

13. Levine A: The p53 tumour suppressor gene and product. Cancer Surveys 12:59-79, 1992.

14. Hunter T, Pines J: Cyclins and cancer. Cell 66: 1071-1074, 1991.

15. Weinberg R: The retinoblastoma gene and gene product. Cancer Surveys 12:43-57, 1992.

16. Almendral JM, Sommer D, MacDonald-Bravo H, et al: Complexity of the early genetic response to growth factors in mouse fibroblasts. Mol. Cell. Biol. 8:2140-2148, 1988.

17. Lord KA, Hoffman-Liebermann B, Liebermann DA: Complexity of the immediate early response of myeloid cells to terminal differentiation and growth arrest includes ICAM-1, Jun-B and histone variants. Oncogene 5:387-396, 1990.

18. Mohn KL, Laz TM, Hsu JC, et al: The immediate-early growth response in regenerating liver and insulin-stimulated H-35 cells: Comparison with serum-stimulated 3T3 cells and identification of 41 novel immediate-early genes. Mol. Cell. Biol. 11:381-390, 1991.

19. Abate C, Curran T: Encounters with fos and jun on the road to AP-1. Sem. Cancer Biol. 1:19-26, 1990.

20. Curran T, Franza BR: Fos and jun: The AP-1 connection. Cell 55:395-397, 1989.

21. Jones N: Transcriptional regulation by dimerization: Two sides to an incestuous relationship. Cell 61:9-11, 1990.

22. Sagar SM, Sharp FF, Curran T: Expression of c-fos protein in brain: Metabolic mapping at the cellular level. Science 240:1328-1331, 1988.

23. Williams S, Evan G, Hunt S: Spinal c-fos induction by sensory stimulation in neonatal rats. Neuroscience Lett. 9:309-314, 1990.

24. Williams S, Evan GI, Hunt SP: Changing patterns of c-fos induction in spinal neurons following thermal cutaneous stimulation in the rat. Neuroscience 36:73-81, 1990.

25. Wisden W, Errington M, Williams S, et al: Differential expression of immediate early genes in the hippocampus and spinal cord. Neuron 4:603-614, 1990.

26. Ingvarsson S: The myc gene family proteins and

their role in transformation and differentiation. Seminars Cancer Biol. 1:359-369, 1990.

27. Kelly K, Cochran BH, Stiles CD, Leder P: Cell specific regulation of the c-myc gene by lymphocyte mitogens and platelet-derived growth factor. Cell 35:603-610, 1983.

28. Hann SR, Thompson CB, Eisenman RN: C-myc oncogene protein synthesis is independent of the cell cycle in human and avian cells. Nature 314:366-369, 1985.

29. Rabbitts PH, Watson JV, Lamond A, et al: Metabolism of c-myc gene products: c-myc mRNA and protein expression in the cell cycle. EMBO J. 4:2009-2015, 1985.

30. Thompson CB, Challoner PB, Neiman PE, Groudine M: Levels of c-myc oncogene mRNA are invariant throughout the cell cycle. Nature 314:363-366, 1985.

31. Waters C, Littlewood T, Hancock D, et al: c-myc protein expression in untransformed fibroblasts. Oncogene 6:101-109, 1991.

32. Dean M, Levine RA, Ran W, et al: Regulation of c-myc transcription and mRNA abundance by serum growth factors and cell contact. J. Biol. Chem. 261:9161-9166, 1986.

33. Moore JP, Hancock DC, Littlewood TD, Evan GI: A sensitive and quantitative enzyme-linked immunosorbence assay for the c-myc and N-myc oncoproteins. Oncogene Res. 2:65-80, 1987.

34. Alitalo K, Koskinen P, Makela TP, et al: Myc oncogenes: Activation and amplification. Biochem. Biophys. Acta 907:1-32, 1987.

35. Penn LJZ, Brooks MW, Laufer EM, Land H: Negative autoregulation of c-myc transcription. EMBO J. 9:1113-1121, 1990.

36. Campisi J, Gray HE, Pardee AB, et al: Cell-cycle control of c-myc but not c-ras expression is lost following chemical transformation. Cell 36:241-247, 1984.

37. Eilers M, Schirm S, Bishop JM: The MYC protein activates transcription of the alpha-prothymosin gene. EMBO J. 10:133-141, 1991.

38. Evan G, Wyllie A, Gilbert C, et al: Induction of apoptosis in fibroblasts by c-myc protein. Cell 63:119-125, 1992.

39. Littlewood T, Amati B, Land H, Evan G: Max and c-Myc/Max DNA binding activities in cell extracts. Oncogene. In press.

40. Heikkila R, Schwab G, Wickstrom E, et al: A c-myc antisense oligodeoxynucleotide inhibits entry into S phase but not progress from G0 to G1. Nature 328:445-449, 1987.

41. Blackwell TK, Kretzner L, Blackwood EM, et al: Sequence-specific DNA binding by the c-Myc protein. Science 250:1149-1151, 1990.

42. Blackwood EM, Eisenman RN: Max: A helix-loop-helix zipper protein that forms a sequence-specific DNA-binding complex with Myc. Science 251:1211-1217, 1991.

43. Vinson C, Garcia K: Molecular model for DNA recognition by the family of basic-helix-loop-helix-zipper proteins. New Biologist 4:396-403, 1992.

44. Land H, Parada LF, Weinberg RA: Tumorigenic conversion of primary embryo fibroblasts requires at least two cooperating oncogenes. Nature 304:596-602, 1983.

45. Stone J, de Lange T, Ramsay G, et al: Definition of regions in human c-myc that are involved in transformation and nuclear localization. Mol. Cell. Biol. 7:1697-1709, 1987.

46. Penn L, Brooks M, Laufer E, et al: Domains of human c-myc protein required for autosuppression and cooperation with ras oncogenes are overlapping. Mol. Cell. Biol. 10:4961-4966, 1990.

47. Freytag SO: Enforced expression of the c-myc oncogene inhibits cell differentiation by precluding entry into a distinct predifferentiation state in G0/G1. Mol. Cell. Biol. 8:1614-1624, 1988.

48. Bursch W, Kleine L, Tenniswood M: The biochemistry of cell death by apoptosis. Biochem. Cell. Biol. 68:1071-1074, 1990.

49. Cohen J, Duke R, Fadok V, KS S: Apoptosis and programmed cell death in immunity. Ann. Rev. Immunol. 10:267-293, 1992.

50. Harmon BV, Takano YS, Winterford CM, Gobe GC: The role of apoptosis in the response of cells and tumours to mild hyperthermia. Int. J. Radiat. Biol. 59:489-501, 1991.

51. McConkey DJ, Orrenius S, Jondal M: Cellular signalling in programmed cell death (apoptosis). Immunol. Today 11:120-121, 1990.

52. Williams GT: Programmed cell death: Apoptosis and oncogenesis. Cell 65:1097-1098, 1991.

53. Wyllie AH: Apoptosis: Cell death in tissue regulation. J. Path. 153:313-316, 1987.

54. Cotter TG, Lennon SV, Glynn JG, Martin SJ: Cell death via apoptosis and its relationship to growth, development and differentiation of both tumour and normal cells. Anticancer Res. 10:1153-1159, 1990.

55. Lennon SV, Martin SJ, Cotter TG: Induction of apoptosis (programmed cell death) in tumour cell lines by widely diverging stimuli. Biochem. Soc. Trans. 18:343-345, 1990.

56. Ellis RE, Horvitz HR: Two C. elegans genes control the programmed deaths of specific cells in the pharynx. Development 112:591-603, 1991.

57. Hockenbery D, Nunez G, Milliman C, et al: BCL-2 is an inner mitochondrial membrane protein that blocks programmed cell death. Nature 348:334-336, 1990.

58. Hockenbery DM, Zutter M, Hickey W, et al: BCL2 protein is topographically restricted in tissues characterized by apoptotic cell death. Proc. Natl. Acad. Sci. USA 88:6961-6965, 1991.
59. Korsmeyer SJ, McDonnell TJ, Nunez G, et al: Bcl-2: B cell life, death and neoplasia. Curr. Top. Microbiol. Immunol. 166:203-207, 1990.
60. Sentman CL, Shutter JR, Hockenbery D, et al: bcl-2 inhibits multiple forms of apoptosis but not negative selection in thymocytes. Cell 67:879-888, 1991.
61. Solomon E, Borrow J, Goddard A: Chromosome aberrations and cancer. Science 254:1153-1160, 1991.

18

ISOLATION OF RARE CELLS BY HIGH-SPEED FLOW CYTOMETRY AND
HIGH-RESOLUTION CELL SORTING FOR SUBSEQUENT MOLECULAR
CHARACTERIZATION - APPLICATIONS IN PRENATAL DIAGNOSIS,
BREAST CANCER AND AUTOLOGOUS BONE MARROW TRANSPLANTATION

James F. Leary, Donald Schmidt, Janet G. Gram, Scott R.
McLaughlin, Camille DallaTorre, Stefan Burde, and Steven
P. Ellis

INTRODUCTION

As we have discussed previously (1), analysis and
isolation of rare cell subpopulations have been of
interest to researchers and clinicians in many areas of
biology and medicine including: a) detection of somatic
cell mutations in mutagenized cells (2), b) detection of
human fetal cells in maternal blood for prenatal diagnosis of birth defects (3), c) detection of CALLA+ cells
(4), d) detection of minimal residual diseases (5,6), e)
detection of stem cells (7), and f) detection of rare
HIV-infected cells in peripheral blood (8). Unfortunately, conventional flow cytometer/cell sorters operating
at rates below 10,000 cells/sec require many hours to
analyze and/or isolate cell subpopulations of low frequencies (e.g. 10^{-4} - 10^{-7}) making them impractical to
use for routine analysis and sorting of such cell subpopulations. One simple method for processing cells at
higher speeds on conventional flow cytometers has been
described (9). This method triggers the data acquisition or sort signal on a rare fluorescence signal.
While useful, this simple method loses the original
frequency information of the rare cells and is blind to
the presence of neighboring non-trigger cells in the
sort unit so that there is no control over sort contamination. A discussion of both theory and practical
implications of "rare-event" theory is given in a recent

book chapter by Leary (10). Despite the limitations of commercially available systems, rare-event analysis techniques have been seen for some time to be of importance for a variety of clinical applications (11). For this reason, despite the arduous nature of the task, many continue to work hard to accomplish their rare cell experiments on conventional flow cytometer/cell sorters.

Attempts to isolate rare cell subpopulations date back to the early 1970's. However, researchers lacked the advanced flow cytometric and sorting technology to practically analyze and separate rare cell subpopulations. For this reason, many investigators continue to use a wide variety of other cell separation methods; for a good review see Kompala and Todd (12). In many instances, these other cell separation methods can and should be used, but in the case of small numbers of difficult to isolate rare cells, one can make strong arguments that flow cytometry/cell sorting is the more appropriate technology. Such an argument will become more obvious as the reader progresses through this paper.

More recently, systems have been built to analyze and/or separate cells at higher rates (13,14,15). High-speed "zap" sorting based on photo-destruction of undesired cells is one way researchers have tried to circumvent the rate limitations of droplet sorting (16). These systems employ newer methods and technological advances such as faster analog-to-digital converters and multi-stage buffering of incoming signals to achieve faster cell processing rates. However, these systems still require digitization and storage of information as correlated or uncorrelated data consisting of all signals from all cells. This digitization of all signals while important and necessary for some applications, particularly for processing of non-rare cell subpopulations, imposes severe and in many cases unnecessary

restrictions on the analysis and sorting of rare cell
subpopulations. In the case of "rare event analysis",
most of the cells are necessarily "not-of-interest". A
simpler alternative is to perform real-time classification of signals as "of interest" or "not of interest"
prior to digitization. This can be thought of as "data
sorting" prior to "physical sorting". Then by using
relatively simple and relatively inexpensive circuitry,
all cells can be counted for original frequency information while digitization is performed only on cells "of
interest" or from cells which cannot be reliably classified by this procedure ("not-sures").

The benefits of such a real-time classification
method are two-fold. First, the circuitry required to
operate flow cytometers at rates of more than 100,000
cells/sec becomes simpler and less expensive and can be
implemented on existing commercially available flow
cytometers. Second, it reduces the problems of storing
and analyzing listmode data sets containing 10^7 - 10^9
cells by storing only listmode data "of interest" or
listmode data about which the experimenter cannot be
certain as to whether it needs to be stored for further
analysis ("not-sures"). Data classified by the system
as "not of further interest" can be counted but not
digitized and/or stored. Over a number of years, a
multiparameter hardware/software system (17) was developed which, when attached to either home-built or commercially available flow cytometer/cell sorters, allows
multiparameter analysis of rare cell subpopulations at
rates in excess of 100,000 total cells/sec. The high-
speed system provides for high speed counting, logic-
gating, count-rate error-checking, and by acting as a
high-speed front-end filter of signals, the system can
be used to control high-speed cell sorting. Use of
thresholds and logical gating from all cells ("total")
and rare cell signals with other total signals allows

multiparameter rare-event listmode data (a record of all
pulses for each rare cell and the total number of all
cells to obtain original frequency information vital to
many applications) to be acquired reasonably both in
terms of signal processing speeds and total amount of
data to be stored by a conventional second-stage data
acquisition system. Analysis of these multiparameter
rare-event data also permit reduction or elimination of
many "false positives", an important problem in the
analysis and isolation of rare cell subpopulations.
This is achieved by further real-time data processing
techniques such as principal component analysis, provided by second-stage processing of original signals as
shown in Fig. 1.

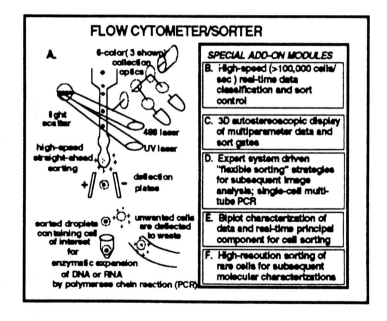

Figure 1. Special high-speed analysis and sorting
system modules (shown on the right) can be linked to
existing commercial and home-built flow cytometer/cell
sorters (shown on the left). The DNA or RNA from rare
sorted cells can be enzymatically amplified by polymerase chain reaction (PCR or RT-PCR) to yield preparative
amounts of DNA or RNA for subsequent dot-spot hybridizations to known sequences.

This paper describes multiparameter, high-speed flow cytometric measurement and cell sorting methods for isolation of rare subpopulations of cells for subsequent molecular characterization by either polymerase chain reaction (PCR) enzymatic expansion of sequence specific DNA or by fluorescence in-situ hybridization (FISH). Using PCR, small numbers (even one!) of sorted cells can be sufficient to generate the equivalent of 10^8 - 10^9 cells for specific sequences of DNA or RNA for which the experimenter has flanking sequence PCR primers (e.g. HLA-DQα primers). FISH on small numbers of cells sorted on the basis of any properties that can be measured by flow cytometry can be used to look directly by fluorescence microscopy or confocal microscopy at copy number of particular chromosomes, chromosome aberrations and rearrangements, and for presence of specific mRNA. Cells can be sorted for subsequent labeling by FISH and visualization by microscopy. FISH-labeled cells can also be detected directly by flow cytometry and sorted on that basis for subsequent image cytometry. Thus the discussion is necessarily very multidisciplinary. Detection, isolation and molecular characterization of rare cells requires the coordination of several different technologies. Rare cells, once isolated, can be examined by a wide array of other powerful technologies, particularly at the molecular level. The "take-home" message for the reader should be that the combination of high-speed flow cytometry/cell sorting, image analysis/ confocal microscopy, FISH and PCR is powerful and can be applied to a wide variety of problems in basic biology and biomedicine.

Some Basic Problems of Rare Cell Sorting

Specific labeling requires that the signal be greater than the "noise". The goal of specific labeling is to improve the signal-to-noise (S/N) ratio to the

point where it permits unambiguous identification of
rare cells for cell sorting. Frequently, no one parameter unequivocally separates the cells of interest. Even
the most highly specific monoclonal antibodies usually
are insufficient for unequivocal identification of rare
cells. Quite often, the number of false-positives is
several times the number of true-positives. However, if
multiple labels are used, the situation often improves.
Each label, while having some amount of non-specificity,
adds new information. For example, use of several monoclonal antibodies labeled with different fluorescent
probes may help to separate a cell subpopulation of
interest from other cell subpopulations in the sample -
something which may not be possible to accomplish with
any one or even two probes. Many false-positive cells
can be eliminated by requiring the presence or absence
of two or more fluorescent signals using either specific
or non-specific antibodies labeled with different fluorescent colors. Addition of other intrinsic flow cytometric parameters such as cell size (by low-angle light
scatter or by light scatter pulse width time-of-flight)
and 90-degree light scatter, can significantly reduce
the number of false-positive cells, thereby improving
the overall S/N of the situation. The problem then
becomes one of an inability to visualize all of the
dimensions of the now very complex multiparameter data.
Methods discussed elsewhere in this paper show how
higher dimensional space can be visualized using principal component analysis and/or use of special 3-dimen-
sional (3-D) computer displays which can be used with a
3-D mouse pointing device to reprocess cells in subregions of this data-space.

Another problem not generally discussed is that of
the statistics that need to be used for some rare cell
applications. For some applications there are so few
cells in the entire sample processed that finite sam-

pling statistics must be used to properly estimate the
true fraction of rare cells (10,18). By using these
sampling statistics, one can determine 95% confidence
limits around an estimate of the rare cell frequency.

A Need for High Speed Signal Processing Electronics

Our early rare cell experiments without the present
high-speed system required nearly 6 hours to run a
positive sample and another 6 hours to run the control
sample. Development of a single-parameter, high-speed,
rare-event analysis system improved the situation and
was successful in some cases in detecting cell subpopulations as low as 0.001% (3). Due to the inherent limitations of detecting a rare cell reliably on the basis
of a single flow cytometric parameter, we subsequently
developed and patented a multiparameter high-speed
system which presently allows selections of rare cells
on the basis of up to three rare parameters (e.g. three
colors of fluorescence) (17). Use of this special high-
speed circuitry as described here allowed multiparameter
analyses of rare cells at rates in excess of 100,000
cells/sec which reduced the time per sample to approximately 15 min. Conventional cell sorters are not designed for high-speed analysis. However, with the
addition of a high-speed signal processing module as
described here, otherwise conventional flow cytometers
can be made to process cells at these rates. The cells
do **NOT** travel faster through the flow cytometer. They
are merely closer together and therefore require better
"electronic book-keeping". The problem is compounded by
the asynchronous arrival of cells. Good signal processing system design must take into account the Poisson
arrival statistics of cells as described by queuing
theory. For a good review of queuing theory see Gross
and Harris (19). For a good theoretical description of
the effects of queuing theory on signal processing

design for flow cytometry see van Rotterdam et al (20).
The total signal processing dead-time of commercial
instruments is typically on the order of 15-50 μsec.
Such deadtimes cause most cells to be missed by the
system when more than 10,000 cells/sec are processed.
The high-speed pre-processing circuitry discussed here
acts as a real-time data classifier for conventional
flow cytometer/cell sorters and reduces effective instrument deadtime to below 2 μsec - roughly the transit
time for the cell to traverse the excitation beam. For
this reason, there is not much point in reducing the
deadtime much below this value. Few cells are missed by
this system at rates in excess of 100,000 cells/sec.
Our experimental data agree well with queuing theory
predictions provided that the cells are not damaged and
sticky and provided that the viscosity of the sample
stream is equivalent or less than that of a solution of
0.25% bovine serum albumin in phosphate buffered saline.
Actual deadtime of the system may vary to be slightly
more than this depending upon both the size of the cells
and the laser beam width, but has been typically less
than 3 μsec for all applications.

Unfortunately, the ability to acquire single parameter data "of interest" previously reported by us (3)
frequently does not discriminate well between true
positive rare cells and false-positive cells from the
major population. Thus an important design philosophy
of our new system is that multiple signals of the rare
cell subpopulation "of interest" as determined by a
suitable "multiparameter rare condition" (and some other
cells which cannot be unequivocally distinguished ["not-
sures"] on the basis of the measured parameters) are
properly gated, digitized and stored in a way which
maintains the correlation between multiple parameters
from both the rare and "not-sure" cells. For many
applications, it is not necessary to digitize tens of

millions of signals which are clearly not of interest.
Not only does this greatly simplify the electronics, but
it would not be practical to store and process tens of
millions of data points in listmode files. Nor do most
data analysis computers have enough random access memory
to have in memory very large listmode files. Most of
the classification (a "positive" or "negative" cell) can
be performed by the very fast analog comparators of this
high-speed module. As much data as desired by the
experimenter belonging to the "uncertain" classification
area can be digitized for further analysis.

A second important feature of our new system is its
use of high-speed gating techniques. The ability to
tightly coordinate signals from the same cells occurring
at slightly different times for a "multiparameter rare
condition" as determined by Boolean logic is important.
This is especially true for an object which possesses
properties that are not uniformly distributed throughout
the object. For example, the nucleus of a biological
cell may be centered in the cell, or positioned near one
of the edges. This system addresses the problem by
monitoring for the desired property (e.g. presence of a
cell's nucleus) throughout some period (e.g. fixed time
period or period defined by some measured physical
property such as the membrane edges of a biological
cell) before making a decision about the object's
"multiparameter rare condition" status.

MATERIALS AND METHODS
Immunofluorescence Staining for Flow Cytometry

Using a whole blood lysis procedure, whereby mature
but not nucleated erythrocytes are lysed (21), we first
lysed 0.2 ml of whole blood from a pregnant woman in the
ninth week of pregnancy, pelleted the cells by centrifugation and then rinsed the pellet with cell buffer
(phosphate buffered saline with bovine serum albumin and

sodium azide) (PBSA). We then followed standard direct
immunofluorescent labeling procedures incubating the
cells with FITC-conjugated anti-CD71 antibody (transferrin receptor, T9, Coulter Electronics) and phycoerythrin
(PE)-labeled anti-glycophorin A antibody (GenTrak) for
30 min at 4°. Unstained samples (for autofluorescence
controls) and singly-stained samples were also prepared
and used for instrument setup and for stain controls.
After rinsing cells in PBSA, samples were fixed in 1%
paraformaldehyde overnight. The cells were then pelleted and permeabilized with 0.1% Triton-X100 in phosphate
buffered saline for 15 min at room temperature. The
cells were then stained with 7-aminoctinomycin D (7AAD)
(10 μM; Sigma Chemical Co.) for a minimum of 30 min, and
kept at 4° in the dark until flow cytometric analysis.

FISH-Staining for Flow Cytometry

We used suspension FISH staining methods for flow
cytometry based on previously published methods (22,23)
with the following modifications: a) intact peripheral
blood mononuclear cells rather than isolated nuclei were
used, and b) single-stranded DNA was produced by exonu-
clease III digestion rather than heat denaturation.
With these modifications, most of the cells stained with
strong FISH signals with little cell clumping. For the
experimental results shown in this paper, a biotinylated
human repeat Y-chromosome probe (pHY2.1) (Amersham) was
subsequently labeled with FITC-conjugated Extravidin
(Sigma Chemical Co., St. Louis) on male, mock, "fetal"
cells. After one round of amplification, the DNA of the
cells was counter-stained with a non-saturating concentration (so that the PI signal does not overwhelm the
FITC signal) of propidium iodide (PI; 1 μg/ml, Sigma
Chemical Co.). High speed flow cytometry was performed
on a mixture of these FISH-labeled cells diluted into
whole blood for a final relative frequency of 0.01%.

Image Cytometry and Cookie Cutter Sorting

A lysed whole blood sample from a donor with a known HLA-DQα type (1.1,4) was fixed in 70% ethanol, pipetted onto a specially coated "cookie-cutter" dish coated with poly-L-lysine, and stained with PI for detection of nucleated cells. Cells were measured on an ACAS 570 image analysis/confocal microscopy system (Meridian Instruments, Okemos, MI). PI fluorescence was detected by argon ion laser excitation at 488 nm and fluorescence detection through a 610 nm long pass filter using the following instrument settings: 200 mW laser power with 20% scan strength and a PMT setting of 60%. Each point was scanned 8 times with signal averaging to reduce background noise. Five PI-positive cells were sorted by "cookie cutting" (24) using 100% laser power at 200 mW, with a cookie radius of 100 μm. The cookies were lifted off the dish under a low magnification (40x total) inverted microscope using a sterile spatula, and individually transferred to PCR tubes for characterization of HLA-DQα alleles using ASO probe analysis as described in the next section. Of the 5 samples, 2 exhibited the correct HLA-DQα type of 1.1,4. The 3 remaining tubes were blank. The most likely cause of blank results is failure to successfully transfer the cell from the ACAS 570 to the PCR tube.

PCR and Allele-Specific Oligonucleotide (ASO) Probe Analysis

Cells used for the molecular characterization by PCR (25) shown in Fig. 13 were sorted directly into PCR tubes containing 40 μl of DNA extraction buffer plus non-ionic detergents. For protein digestion, 1 μl of proteinase-K (from a stock solution at 10 mg/ml) was added to each tube and incubated at 55° for 1 h. The proteinase-K was subsequently inactivated by heating the mixture to 95° for 10 min. Then, 50 μl of HLA-DQα PCR reagents from an Amplitype[TM] kit (Roche Molecular Systems) were added directly to each tube. A "hot-start" (26) was used whereby Mg^{++} was added at the first annealing step after denaturation. Forty cycles of PCR amplification were then performed. At the end of the PCR amplifications, the PCR-amplified DNA sequences from the sorted cells were characterized by dot blotting to ASO probes to these DNA sequences using AmpliType[TM] kit.

Multiplex PCR

One way to perform useful molecular diagnostics is to simultaneously use multiple sets of PCR primers to confirm cell identity and/or check for a gene defect of interest in a "multiplex PCR" reaction. An example of this methodology is shown in Fig. 15. If insufficient DNA is available for a multiplex reaction, an alternative is to perform "whole genomic amplification", whereby approximately 30 genome equivalents of DNA can be obtained from a single cell by alu-PCR (27), degenerate oligonucleotide primed PCR (DOP-PCR) (28), primer-extension-preamplification (PEP-PCR) (29) or other genomic amplification techniques. In general, it is probably preferable to perform a whole genomic amplification step first. PCR is difficult to perform reliably and reproducibly on single cells, but is relatively straightforward at the level of 5-cell equivalents or more. As demonstrated here, it is **NOT** desirable to actually perform PCR on 2 or more sorted rare cells because there is always the possibility of one or more of the sorted cells being a false-positive cell (cf. discussion of single cell, multi-tube PCR).

FLOW CYTOMETRY AND CELL SORTING
General Description Of The High Speed Flow Cytometry/ Cell Sorting System

The following represents a description of the

method and apparatus of the HiReCS (High-Resolution Cell Sorter). A functional block diagram of the high speed system, including the high speed circuitry and its interconnections to a flow cytometer/cell sorter is given in Fig. 2.

A critical feature of this system is the ability to determine multiple rare signals, or lack thereof, from a cell at very high signal processing rates, and to rapidly determine a Boolean logic condition of these multiple rare signals as well as correlated total (other parameters similar to those on non-rare cells) parameters. This allows detection of rare cells on the basis of multiple criteria according to any Boolean combination

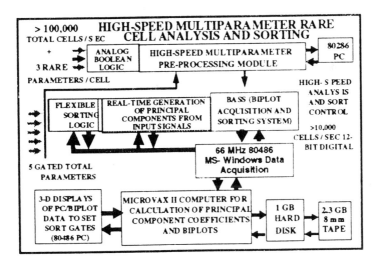

Figure 2. The high-speed module acts as a high-speed front-end signal processing filter allowing conventional flow cytometer/cell sorters to operate at rates in excess of 100,000 cells/sec. Real-time data classification of cells into 3 categories: positive, negative and "not-sure" can occur at rates in exess of 100,000 cells/sec. Only signals from "positive" and "not-sure" cells are passed along to the second stage principal component/biplot module which attempts to classify the "not-sure" cells into "true-positives" and "false-positives". High speed lookup tables (LUT's) are used to transform input signals in real-time to create principal component parameters and to permit "flexible sorting" of cells from principal component space.

of rare parameters (the "multiple rare condition") including lack of a given rare signal. Thus, there is considerable flexibility in the choice of which cells will have their correlated parameters digitized.

A sophisticated real-time data control program for the system called "HISPEDPC" has been developed in our laboratory. This software provides the operator with a flexible, menu-driven graphical user interface to the high-speed hardware while at the same time providing very precise control over the process of data acceptance. The program is currently written in Borland Turbo C and currently runs on an IBM-compatible PC under MS-DOS. The menu provides the operator with a set of commands for system setup and control, rate window specification, and data display and storage. The menu portion of the screen consists of pull-down menus and an area to allow system prompts and other messages to be displayed while the count and/or percent rare data are continually updated. The operator can manually stop the data acquisition at any time. At the menu level, current counts and rate histograms may be separately viewed, saved or purged; the system may be configured or re-configured, if so desired, using saved setup files. An example of the HISPEDPC display from a typical run is shown in Fig. 3. The rate checking boundaries shown in this example would be very sensitive to small fluctuations in sample arrival rate.

Very sophisticated "multiparameter rare conditions" can be set using combinations of Boolean logic. The RARE1/RARE2/RARE3 condition on each line constitutes a logical **AND** situation. By combining multiple Boolean logic lines virtually any logic situation can be defined by the user. A second feature of importance in this portion of the high-speed data acquisition system is a burst-rate histogram overlaid on the cumulative rate histogram that allows for real-time detection of unde-

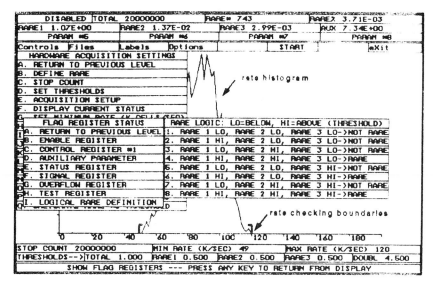

Figure 3. High speed system user-interface with drop-down menus. Rate histograms, when used for event burst-mode rejection by error checking software, allow high-speed process control for both data acquisition and sorting. Complex "multiparameter rare conditions" can be defined in terms of Boolean logic conditions.

sired "event bursts" whereby cells arrive at abnormal rates (typically bunched very closely together). Burst mode rejection has been used sucessfully by others to remove false-positives from rare cell listmode data after data acquisition (30), but such burst mode rejection must be done in hardware in real-time to remove false-positives when performing cell sorting. A cumulative rate histogram is shown in Fig. 3 whereby the x-axis is the rate at which cells are being analyzed by the system, in this case approximately 85,000 cells/sec. Since the cells are being processed at too great a speed for even a very alert human operator with reaction time of approximately 0.1 sec to respond to a problem situation (8,500 cells can end up in the wrong place in 0.1 sec!), the rate histogram serves as a very sensitive process control parameter. If flow conditions are proper, the cells should arrive in random arrival for small

time intervals (e.g. 1 msec). The user can define a
small time interval over which a mean cell rate must be
within certain defined bounds. These boundaries can
then be used to detect flow abnormalities so that
the error-checking software can automatically shut down
data acquisition and sorting. Rate changes are a measure of the Poisson arrival statistics of cells and are
a sensitive indicator for potential sorting problems.

Another critical feature of the rest of the data
acquisition system described elsewhere in this paper is
that complex multiparameter data can not only be viewed
using principal components but, in fact, sorted on the
basis of principal components. Using special hardware
and software LUT, we can calculate principal components
for each cell in real-time between the laser excitation
point and the sort decision point. Thus the system is
capable of sorting out cell subpopulations that are not
even visualizable, in some cases, by looking at all
possible bivariate displays of the data (cf. section on
visualization of multidimensional data).

Sorting Speed Versus Purity
The sorting of rare cells in droplet sorting systems is ultimately limited by the number of droplets/
sec. This problem is a straight-forward application of
queuing theory; if cells arrive randomly, they are
distributed among the subsequently-formed droplets
according to a Poisson distribution. At typical sorting
rates of 2,000 cells/sec, the probability of two cells
occurring inside the same droplet ("coincidence") is
negligible but at higher rates the coincidence rate
sharply increases. Sorting cells of high purity (>95%)
at rates of 5,000 cells/sec or greater requires "anti-
coincidence" circuitry which typically rejects all cells
which are close enough to be sorted in the same sorting
unit (there are typically one, two, or three droplets in

each sorting unit). Most sorting is done with 2 or 3 droplets per sorting unit. Three droplet sorting is very forgiving in terms of setting sort delay times, but necessarily increases the probability of coincidence of two or more cells within the sort unit. One-droplet sorting while very good with respect to lowest probability for coincidence, is very difficult to perform in terms of both setting the proper sort delay time and of maintaining that level of sort stability. While we have been able to perform good one-droplet sorting in some high-speed sorting experiments, we use two-droplet sorting for most experiments.

Sometimes one can tolerate coincidence of some contaminating cells in the sorting interval. A common example for us is the contamination of the interval with a non-nucleated red blood cell. When we are sorting for PCR expansion of DNA from our cell of interest, the presence of small numbers of non-nucleated red blood cells does not appear to disturb PCR reactions. If one is interested only in enrichment rather than absolute purity, at very high sorting rates (e.g. >100,000 cells/sec), there will be multiple cells in each sorted droplet (Fig. 4). While this is generally undesirable, the actual enrichment factor by sorting is greater at higher sorting rates than at lower rates for the case of rare (<0.1%) cell subpopulations. For example, the enrichment factor for cell separation of a 10% cell subpopulation to become >95% pure is less than 10. However, for a high-speed separation of a 0.01% cell subpopulation at 120,000 cells/sec with a droplet rate of 30,000 droplets/sec, the average number of cells in each sorted droplet is 4. Importantly, one (and very rarely more than one) of these four cells will be a positive cell from the 0.01% rare cell subpopulation. Since the original frequency of rare cells is 0.01% and the final frequency is very slightly more than 25%, the

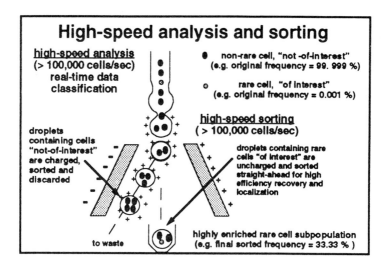

Figure 4. High speed sorting results in several cells occurring within the sorting unit (of one, two, or three droplets). As shown in the example, high-speed cell sorting of rare cells can provide even higher enrichment factors than conventional sorting of non-rare cells (30,000-fold enrichment in the above example).

enrichment factor is more than 2,500. Similarly, for separation of a 0.001% cell subpopulation, the enrichment factor is more than 25,000. Part of the system involves "flexible sorting strategies" (31) designed to allow the experimenter to select a trade-off strategy most suitable for the goals of the experiment including choice of an optimal first "enrichment" sort followed by a second sort either by the cell sorter, by "cookie-cutter" sorting (24), or by optical trapping (32) of cells on a fluorescence image analysis system to obtain cells of the desired level of purity. This "flexible sorting" system shown in Fig. 5 also allows for very complicated sorting algorithms to be used including ones based on the use of other non-flow-cytometric information known about the sample.

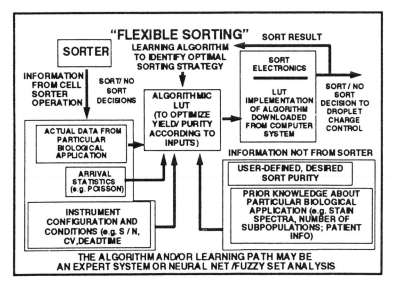

Figure 5. Complex "flexible sorting" strategies and algorithms can be loaded into special sort logic lookup tables (LUT's) allowing sort decisions to be made on the basis of both flow cytometric parameters and other information known about the cells. These strategies can be especially useful when combined with secondary stage micromanipulation of cells of interest from impure sorts.

Visualization of Multidimensional Rare-Cell Data Using Principal Components

By collecting multiple measurements per cell to distinguish rare cell subpopulations, we solve one problem and create another one. As long as each parameter helps enrich for the rare cell of interest, the greater the number of parameters measured the more likely one will be able to isolate the rare cells of interest. Unfortunately, these added flow cytometric parameters produce very complex data that is difficult to visualize by conventional flow cytometric data analysis methods. In addition, the "tyranny of dimensionality" means that data points tend to spread ever more widely in the additional dimensions so that it becomes difficult to visualize any pattern to the data.

Many do not realize that viewing all possible

bivariate views of multiparameter data does **NOT** necessarily allow them to see all of the data. It is possible to completely miss seeing cell subpopulations which are projected down on top of other cell subpopulations. Most people are familiar with the problem that two subpopulations can be projected down onto each other in histograms. Indeed, one of the reasons people commonly use bivariate displays is to avoid this problem. Unfortunately, the same problem can occur when higher dimensional data is projected down onto bivariate displays as shown in Fig. 6.

What we would like to do is to view multiparameter flow cytometric data in as high a dimensionality as possible. But even to view a third dimension to the of data can represent a problem. For humans to visualize 3-D data on a 2-D computer screen or paper we must use

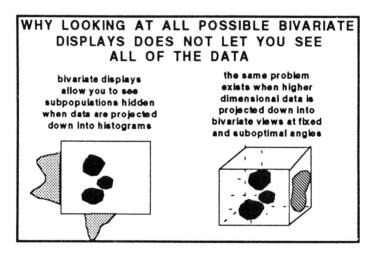

Figure 6. When bivariate data are projected down into their component histograms, discrimination between cell subpopulations is frequently lost. Unfortunately, the same problem exists for higher dimensional data projected down onto a series of bivariate displays. Important information may be lost or not visible even if all possible bivariate displays are viewed. One way to help maximize differences in the data is to look at the first three principal components of the multidimensional data (cf. Fig. 7).

one or more visual cues or tricks to create the illusion depth for the third dimension. One common method, "stereo-pairs", can be used to allow viewing of 3-D data, but it requires the viewer to "fuse" the stereo pairs. Many people find this "fusing" difficult and a small but significant portion of the population actually lack stereoscopic fusing capabilitiy. To make such viewing easier, "autostereoscopic" displays which do not require the viewer to fuse stereo pairs can be used to allow direct viewing of 3-D data on a PC. We now use an autostereoscopic display (Dimension Technologies, Inc., Rochester, NY) to view both 3-D flow cytometry and confocal microscopy data. One problem with these approaches is that as the flow cytometric data is spread in three dimensions, the number of cells in each array element of the display in many cases becomes very small unless very large data sets are analyzed. The problem becomes even more severe as the number of dimensions increases.

Dimensions higher than three can not be directly observed. Unfortunately, successful isolation of rare cell subpopulations frequently requires four or more parameters to be measured on each cell. 4-D data are frequently viewed by looking at all non-redundant bivariate combinations of the data. For example, 4-D flow cytometric data with parameters A, B, C and D would have 6 possible non-redundant bivariate views of the data: A vs B, A vs C, A vs D, B vs C, B vs D, and C vs D (where B vs A, etc. are considered redundant displays). For six-parameter data, there are 15 non-redundant bivariate displays. For eight-parameter data, there are 28 non-redundant bivarate displays, and so on. Not only is it difficult to ask humans to attempt to reconstruct higher dimensional data from these visual "slices" of the data, but all of the slices do not necessarily portray the information in the data since the data will be oriented

arbitrarily in higher dimensional space. Such slices
will tend to project subpopulations down on top of each
other. Clearly, there is a need to find a way of visualizing multidimensional data in a single or a few
optimal displays.

One of the ways higher dimensional data can be
visualized is through the use of principal component
analysis, a special case of the more general "projection
pursuit analysis" (33). Use of the first two principal
components chooses a coordinate system that yields a 2-D
graphical representation of multidimensional data in a
way that shows the greatest variance in the data (Fig.
7).

However, this means that the first two principal
components will tend to be determined by larger cell
subpopulations. Other cell subpopulations, particularly
smaller ones and very rare ones, may not be well-repre-
sented in this choice of principal components or projection unless they lie far from the centroid of the data.
Thus one problem in using principal component analysis
in the detection of rare cell subpopulations is that the
principal component plane will be determined by the
major cell subpopulations which by definition are not
the cell subpopulations of interest. One way to deal
with this problem is to first use gating on original or
principal component parameters to strip away additional
cells "not of interest". The remaining cells can then
be "reprojected" onto a new principal component plane
chosen by the remaining cells. A second problem is that
to find rare cells one must compare a given sample to
its control sample. Since each data set will choose a
different set of principal components to compare two
data sets requires reprojecting both data sets to a
common projection.

Principal component analysis should not be used in
"brute force" fashion. Careful application of principal

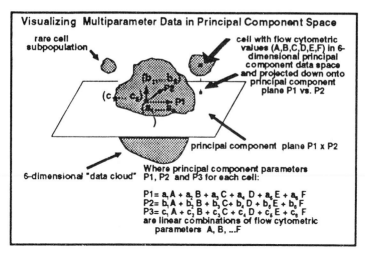

Figure 7. Principal conponent analysis projects higher
dimensional data down onto a two (or three) dimensional
hyperplane which maximizes the variance of the data. In
this way, clusters of higher dimensional data corresponding to cell subpopulations can be visualized by
human observers. Since rare cells may be hidden by the
initial principal component projection, it may be necessary to strip away some of the cells not of interest and
to reproject the data onto a better projection which
unlike the conventional bivariate displays can be controlled to permit optimal viewing of the data.

component analysis can be used to optimize representa-

tion of particular cell subpopulations. One should

consider the use of principal components only as a

"rough first cut" through the multidimensional data-

space. The larger cell subpopulations which may not be

of interest can then be removed from the data set by

gating and reprocessing the data on the basis of princi-

pal components and/or raw parameters. New principal

components can then be calculated on this subset of

correlated, reprocessed listmode data. Such gating,

reprocessing and reprojection of the data onto a new

principal component plane may reveal other cell subpopu-

lations hidden by the original principal component

projections which were chosen to maximize the variance

of the major (rather than the minor) cell subpopulati-

ons.

Principal components are linear combinations of the original flow cytometric parameters. The first principal component P_1 is that linear combination of flow cytometric parameters that gives the maximum spread (variance) of the data points. The second principal component P_2 is the linear combination which gives the next greatest spread of data, and so on for the third principal component P_3 and other principal components. For example,

$$P_1 = a_1 A + a_2 B + a_3 C + a_4 D + a_5 E + a_6 F$$
$$P_2 = b_2 A + b_2 B + b_3 C + b_4 D + b_5 E + b_6 F$$
$$P_3 = c_1 A + c_2 B + c_3 C + c_4 D + c_5 E + c_6 F$$

where $a_1, \ldots b_1, \ldots,$ and $c_1, \ldots$ represent the calculated coefficients for principal components P_1, P_2 and P_3, respectively, and A, B, ... F represent flow cytometric parameters such as green fluorescence, red fluorescence, forward light scatter, 90-degree light scatter, etc.

"Biplots" (34), a technique that combines principal component analysis with a graphical representation of the variance and parameter correlation of the data, can be very useful for determining the probable identity of rare cell subpopulations before sorting and also for reducing false positives. Each subpopulation of cells tends to have a unique biplot pattern that is a unique signature for that cell subpopulation (1).

Sometimes viewing the first two principal components of a data set still causes us to miss cell subpopulations hidden behind other cell subpopulations. We now routinely view the first three principal components. This means that we must now view 3-D trivariate flow cytometric data. Unfortunately, to produce 3-D displays of data for publication we are forced to produce standard "stereo pairs" that many people have trouble viewing (Fig. 8). In our laboratory, we use a special 3-D "autostereoscopic" display system which produces stereo pairs displaced one pixel apart on a display system

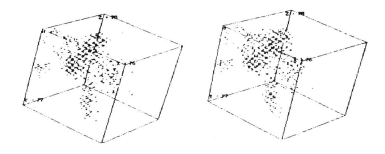

LEFT VIEW   **RIGHT VIEW**

Figure 8. Typically visualization of 3-D data requires visual tricks such as "stereo pairs". Successful visualization requires the observer to "fuse" a third image between the left and right stereo views. Unfortunately, this is presently the only practical way to publish 3-D images. Such limitations and requirements of the observer to fuse stereo images can be overcome by using 3-D computer displays such as the one described in Fig. 9.

(Fig. 9). The result visually is a "hologram-like" image that appears to come out of the display in a third dimension and unlike standard stereo pairs it is easy to visualize.

The autostereoscopic display works as shown in Fig. 10. A vertical slit mask allows each eye to see one and only one of the stereo views. By using the natural parallax of the eyes we can now see 3-D images without requiring the "fusion" of stereo pairs that is difficult for many people. We have written software to allow for routine 3-D viewing of flow cytometry data files in standard FCS format from different flow cytometers. In addition, we have made use of a 3-D mouse pointing device (a standard mouse with the third dimension obtained by holding down one of the mouse buttons) to address subregions of the 3-D data. Listmode data can be reprocessed in 3-D space by drawing a box around the subregion of interest and saving this subset of the original data to a new listmode file. A more elegant

Figure 9. Three-dimensional data (or combinations of original data and Principal Components data) can be conveniently viewed using an autostereo display system linked to a personal computer. The display requires no special glasses for the viewer to see a hologram-like image which can be addressed using a 3-D mouse. Unlike conventional stereo pairs, no requirement is made for the observer to "fuse" the two images since the left eye sees only the left image and the right eye sees only the right image. We are also using this system to view and address 3-D confocal microscopy data. Animation can help the experimenter choose an optimal viewing angle.

method we have developed is to point to the approximate center of each "data-cloud" in the 3-D display and to use it for a starting point to determine the proper number and location of cell subpopulations in a "guided cluster analysis".

An Example of High-Speed, Principal Component/Biplot Analysis of a Rare Cell Subpopulation Stained by FISH

To illustrate all of the preceding discussion of high-speed, principal component/biplot analysis we show in Figs. 11 and 12 data obtained on a defined cell mixture of FISH male (mock "fetal" cells) and unlabeled female (mock "maternal") lymphocytes in a ratio of 1:10,000. This rare (0.01 percent) population of male

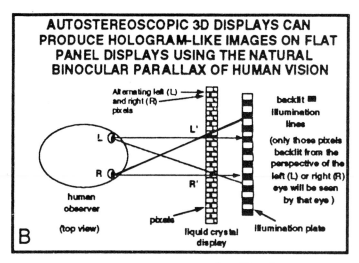

Figure 10. The autostereoscopic display system works by producing left and right stereo images one pixel apart on a special viewing screen. A vertical mask allows the left eye to view only the left stereo image and the right eye to view only the right stereo image for an appropriately situated observer. The mouse cursor is also produced in both left and right stereo images so that it appears at a certain depth within the screen. By holding down on the left button of a conventional mouse, the observer can move in and out of the screen in the third dimension. The mouse can be used to point to approximate cluster centers of cell subpopulations for subsequent "guided cluster analysis" or can, if the right button is depressed, stretch a 3-D rubber-band box around a cell subpopulation of interest for listmode data processing or for the setting of sort gates.

peripheral blood mononuclear cells labeled by FISH were subsequently analyzed through the HiReCS system at a rate of 50,000 total cells/sec. The purpose of these studies was to determine whether fetal cells could be isolated by cell sorting from maternal blood on the basis of the number (e.g. trisomy 18 or 21) or type (Y-chromosome-specific male fetal cells) of chromosomes as labeled by FISH. In Fig. 11, we show photomicrographs of male cells labeled by FISH with a molecular probe which paints the long arm of the human Y-chromosome (and consequently produces a fluorescent spot within the interphase nucleus of the cells we examined by high-speed flow cytometry). The cells were labeled by FISH as described in Materials and Methods and contained a round of amplification with biotinylated anti-avidin and then more fluorescent avidin as shown in Fig. 11B. All cells were also counter-stained with PI (at less than saturating dye concentrations to prevent overwhelming of the FITC signal) and gated on the G_0/G_1 peak to eliminate debris, damaged nuclei, and cell doublets. FISH was performed on a mixture of male human lymphocytes which were then added to a much larger population of female human lymphocytes to a final mixture frequency of 0.01 percent. A biotinylated human repeat Y-chromosome specific probe was labeled with FITC conjugated Extravidin^{TM} (Sigma Chemical Co.) and amplified once using biotinylated anti-avidin antibody. Fig. 12 shows the multiparameter correlated listmode data that was obtained on the rare cells passed through the HISPEDPC system for subsequent digitization and analysis by principal component analysis. The first three principal components were calculated on a cell-by-cell basis as previously described and subsequently added as correlated listmode parameters 7, 8 and 9. These principal component parameters can also be calculated on a cell-by-cell basis in real-time once the principal component coefficients have been calculated for an aliquot of the sample. This now permits real-time cell sorting on the basis of principal component parameters calculated for each cell between the laser intersection point and the sort decision point.

While the separation of positive from negative cells is still difficult due to wide variations in FISH staining, we hope to tighten these distributions by using ratiometric measurements with a FISH probe against the human X-chromosome. Variations in FISH staining appear to be due to variations in the ability of the

A

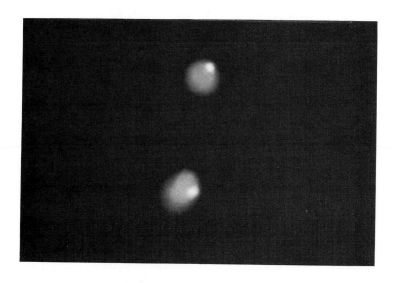

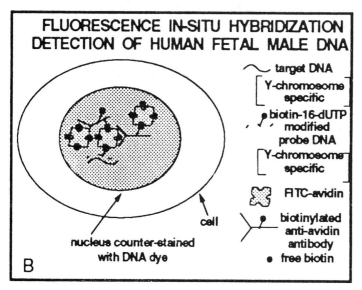

Figure 11. (A) Photomicrograph of male human interphase cells stained by FISH with a molecular probe which paints the long arm of the human Y-chromosome and appears as a spot within cells counter-stained with PI. (B) The Y-chromosome FISH staining was further amplified with a round of biotinylated anti-avidin then more fluorescent (FITC) avidin as shown in this photomicrograph.

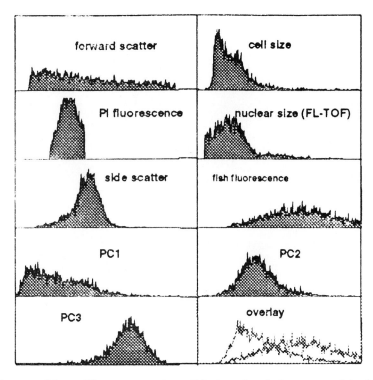

Figure 12. Six-parameter listmode data was obtained by high-speed analysis using the special analysis and sorting system described previously in Fig. 1. A rare (0.01%) population of male human lymphocytes was labeled by suspension FISH with a Y-chromosome-specific probe. Principal components were calculated on a cell-by-cell basis and subsequently included as correlated listmode parameters 7, 8 and 9. The overlay panel compares FITC fluorescence of a sample with FISH-stained male lymphocytes doped in at a frequency of 0.01% with a control sample containing male cells stained similarly but with no probe present at the first step.

probe to have access to their DNA targets. By using a second reference probe, e.g. against the X-chromosome, the distributions tend to tighten (35). Thus, our strategy will be to perform high-speed dual FISH staining, to isolate the putative fetal cells by high-speed sorting, and to confirm by fluorescence microscopy and/or confocal microscopy.

One of the problems in examining fetal cells with FISH probes is that they are so rare in maternal blood that it is extremely difficult to find enough of them to analyze. By using high-speed cell sorting, even if we cannot isolate the FISH stained cells with absolute purity, we can enrich rare cell subpopulations to the point where subsequent fluorescence microscopy, image analysis or even confocal microscopy becomes not only possible but practical. Thus combining methodologies as shown in Fig. 13, we can perform many forms of secondary cell and data processing on rare cell subpopulations.

PCR Expansion of Specific DNA or RNA From Rare Sorted Cells

The advent of new technologies outside the field of cell separation can have major impact on the practicality of some cell separation techniques. Until recently, cell sorting has been confined mainly either to the cloning of single or small numbers of cells for growth in tissue culture or for subsequent assays which require only small numbers of cells. Cell sorting cannot sort

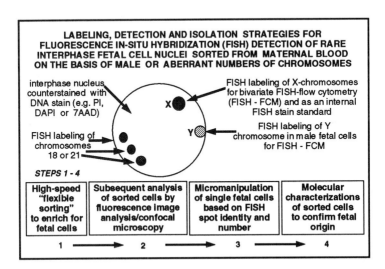

Figure 13. When rare cells are sorted even if only into enriched subpopulations, it becomes practical to perform secondary processing of these cells by a wide array of other methods and technologies.

preparative amounts of cellular material for traditional
biochemical analysis. However, PCR changes this picture
drastically in terms of DNA or RNA which can be enzymatically expanded rapidly in vitro. Thus, a very small
number of sorted cells (even a single cell) can have
portions of its DNA or RNA enzymatically expanded to
produce the equivalent amount of this DNA or RNA as if
one had sorted 10^6 to 10^9 cells. Subsequently, standard
molecular biology techniques can be used on the PCR-
expanded DNA or RNA from small numbers or even a single
rare sorted cell. The trick becomes to accurately sort
the correct cell, i.e. to perform "high-resolution"
sorting of rare cells. Unfortunately, it is frequently
not possible to obtain cell markers which can unequivocally detect the rare cells of interest. Later, we will
describe a new method of isolating pure rare cells. In
this method, we solve the problem of the inability of
perfect marker identification of the rare cells by
"single-cell, multi-tube sorting" for PCR.

An Application of Analysis of Rare Human Fetal Cells from Lysed Maternal Blood

Another rare-cell application involves detection of
human fetal cells in maternal blood on the basis of
differentiation antigens. Many of the fetal cells in
the maternal blood are nucleated erythroid lineage cells
bearing differentiation antigens expressed early in
erythroid development. An example of this is shown in
Fig. 14A which shows the 3-color detection of rare human
fetal nucleated erythroid cells from the maternal blood
of a woman in the ninth week of pregnancy on the basis
of FITC-anti-transferrin (CD71) monoclonal antibodies,
PE-glycophorin-A monoclonal antibodies, and 7-amino-
actinomycin D (7AAD) which stains the nucleus of all
cells. The gated regions of the histograms shown in
Fig. 14B were used to sort cells for subsequent molec-

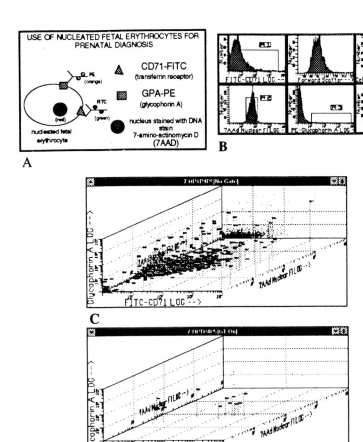

Figure 14. (A) Nucleated fetal cells of early erythroid lineage can be identified on the basis of developmental markers labeled by fluorescent monoclonal antibodies and flow cytometry. Since there will be a large number of maternal non-nucleated erythroid cells (e.g. reticulocytes) containing hemoglobin which can interfere with the PCR reaction, it is wise to also stain for DNA using 7-aminoactinomycin D.
(B) Histograms showing cell sorting gate regions.
(C) A trivariate histogram of ungated data.
(D) Using this strategy, fetal cells can be identified at frequencies of approximately 10^{-6} in maternal blood and sorted using the sort windows (Boolean AND of regions R1+R2+R3+R4+R5) are shown in the gated trivariate display of Fig. 14D, a subset (10 cells) of which were sorted for subsequent molecular characterization (Figs. 15 and 16).

ular analyses shown in Figs. 15 and 16. An ungated trivariate display (using WinList, Verity Software House, Topsham, ME) shows considerable heterogeneity in the data. Upon gating on regions 1-5 of Fig. 14B, 10 cells were sorted from the region as shown by the trivariate display of Fig. 14C. Unfortunately, it is possible for the pregnant mother to also express these antigens. Hence the markers do not permit unequivocal detection of fetal cells as shown in the next section (cf. Fig. 16) in a sort of 10 cells which upon subsequent molecular analysis revealed a mixture of one or more fetal cells amid a total of 10 sorted cells which obviously contains maternal cells.

Molecular Characterization of Sorted Rare Cells

One of our research interests is the high-resolution sorting of rare human fetal cells from maternal blood for a risk-free, non-invasive prenatal diagnosis of birth defects. Fetal cells can be sorted from maternal blood by the techniques just described. Sorted fetal cells can have portions of their DNA enzymatically amplified by PCR for further genetic analysis using dot-spot hybridizations to ASO (Fig. 15A). PCR primer sequences used in the Amplitype^{TM} kit (by Roche Molecular Systems through Perkin Elmer, Inc., Norwalk, CT) to amplify HLA-DQα sequences in sorted cells are shown in Fig. 15B. Actual data from human fetal cells obtained by sorting 10 cells from mother's blood and applying ASO analysis using the DQα alleles in the HLA gene locus amplified by PCR using Amplitype^{TM} kits is shown in Fig. 16. Having previously determined the HLA-DQα types of the mother (1.2,1.3) and the father (1.1,2), the fetus must share one allele from each parent. The 10 cells sorted from the mother's blood before delivery express both maternal and paternal alleles (1.1,1.2,1.3) indicating that the 10 sorted cells represent a mixture of

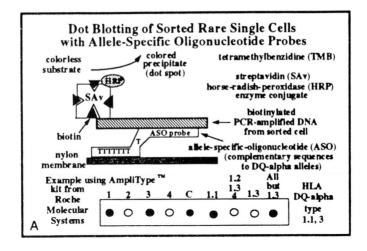

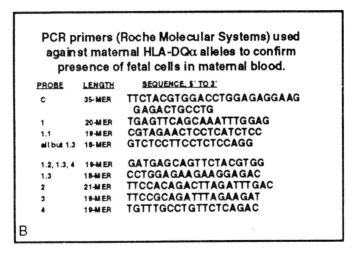

Figure 15. (A) Concept of allele specific oligonucleotide (ASO) dot spot hybridization of DNA sequences PCR
amplified from rare sorted fetal cells.
(B) PCR primer sequences used to amplify HLA-DQα DNA
sequences in the sorted cells.

maternal and fetal cells. Upon birth the baby was then
HLA-DQα typed as (1.1,1.2) proving that at least one
cell among the 10 cells sorted was fetal because the
HLA-DQα 1.2 paternal allele was present.

There are at least two important advantages of
using paternal HLA gene sequences to detect fetal cells.
First, the fetus can be either male or female. Second,

306

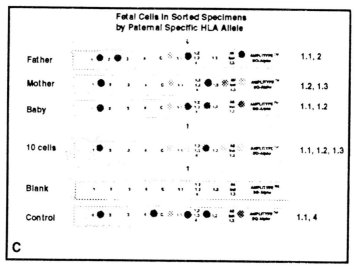

Figure 16. ASO dot spot hybridization of PCR-amplified DNA sequences from cheek epithelial from a father, mother and baby. Before the baby's birth, fetal cells were isolated from the mother's blood by high-speed 3-color fluorescence analysis as described in Fig. 11. Note that the 10 sorted putative fetal cells contain both paternal and maternal HLA-DQα alleles indicating that the sorted cells contain a mixture of fetal and maternal cells, demonstrating the need for single-cell, multi-tube PCR analysis (cf. Figs. 17 and 18).

sources of contamination (a problem when Y-specific sequences are used as primers) can be identified. When combined with other paternal-specific markers such as group component (Gc) primers, the identification of fetal cells is unequivocal. Obviously, this combination of markers is unable to completely distinguish fetal from maternal cells. However, it is good enough to enrich the fetal cells from an original frequency of less than 1 fetal cell per million maternal cells to at least 10 percent, a 100,000-fold enrichment on a single pass. Other enrichment techniques generally lead to higher losses, particularly if there are only a very small number of rare cells present, and would lose original frequency information. A solution to this problem is given in the next section.

Single-Cell, Multi-Tube Sorting to Obtain Pure Rare Cells

To obtain pure rare fetal cells we are now attempting to perform "single-cell, multi-tube PCR" whereby one cell is sorted directly into each of a number of PCR tubes containing DNA extraction buffer plus non-ionic detergents (Fig. 17A). In Fig. 17B, we show results from single cells sorted by this single-sorted cell/ multi-tube method and subsequently amplified by PCR using HLA-DQα primers. Fig. 18 shows results obtained from defined adult mock mixtures of "fetal" and "maternal" cells of known HLA-DQα types and the success rates for one or more "fetal" cells either pure or mixed with contacell is sorted directly into each of a number of PCR tubes containing DNA extraction buffer plus non-ionic detergents (Fig. 17A). In Fig. 17B, we show results from single cells sorted by this single-sorted cell/multi-tube method and subsequently amplified by PCR using HLA-DQα primers. Fig. 18 shows results obtained from defined adult mock mixtures of "fetal" and "maternal" cells of known HLA-DQα types and the success rates for one or more "fetal" cells either pure or mixed with contadefect. A number of strategies for multiplex PCR analyses are shown in Fig. 19. The HLA-DQα plus Gc variant results prove that the cells are fetal and not maternal. A third set of primers can simultaneously confirm the specific gene defect. Several multiplex PCR methodologies are shown in Fig. 20 whereby DNA sequences in the HLA-DQα gene region and the Gc variant region have been simultaneously expanded by PCR. However, performing PCR on single cells can be difficult and sometimes leads to failure to PCR amplify all primer regions adequately leading to "allelic loss". If insufficient DNA is available for multiplex PCR, genomic DNA amplification methods can be performed to yield approximately 30 copies of DNA from a single cell (29). Then aliquots of this DNA can be used for individual reactions which are more reliable and reproducible than multiplex reactions performed on single cells.

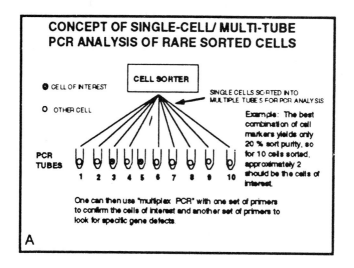

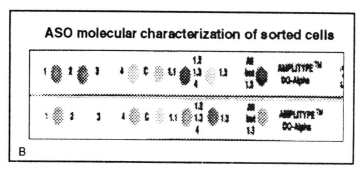

Figure 17. (A) Concept of single-cell/multi-tube sort and PCR analysis of rare cells for which no combination of markers directly yields cell subpopulations of 100% purity. If the sort is only 30% pure then only 3 of the 10 tubes will be likely to contain the rare fetal cells of interest. However, these three tubes contain 100% pure fetal cells which can be subsequently analyzed for gene defects anywhere on the fetal genome by multiplex PCR.
(B) Individual cells can be molecularly characterized using Amplitype[TM] kits. When 1 mock "fetal" cell is sorted with 19 mock "maternal" cells into one tube, molecular characterization shows both "fetal" and "maternal" alleles.

MULTI-TUBE, SINGLE CELL SORT / PCR RESULTS

CELLS SORTED per TUBE	"FETAL"			"MATERNAL OR INCONCLUSIVE"			FETAL DETECTION EFFICIENCY
	P+S+M+	P+S+	P+	S+M+	S+	blank	
1 FETAL (n=10)	-	6	1	-	1	2	7/10
10 FETAL (n=10)	1*	9	-	-	-	-	10/10
1 FETAL 19 MATERNAL (n=10)	6	-	-	2	1	1	6/10
3 FETAL 57 MATERNAL (n=10)	10	-	-	-	-	-	10/10
57 MATERNAL (n=9)	-	-	-	9	-	-	N/A

HLA-DQα GENOTYPES:
Fetal Genotype: [1.2,4]
Maternal Genotype: [1.2,2]
*#2 allele may be due to coincident sorting of a maternal cell or laboratory contamination

DETECTED ALLELES:
P+: Paternal allele - 4
S+: Shared allele - 1.2
M+: Maternal allele - 2

C

Figure 18. Efficiency of ASO molecular characterization of sorted mock "fetal" and "maternal" cells on the basis of cell number and in the presence of other contaminating cell types. Under certain conditions of fixation and staining, ASO characterization can be performed at the single-cell level. Even presence of contaminating cells at the 95% contamination level does not interfere with successful detection of the fetal cell alleles provided that they differ from that of the mother in at least one allele. If the allele of interest is shared by the mother then we must sort to single cell level to insure proper confirmation of the status of that fetal cell allele.

Recent Applications in Breast Cancer and Bone Marrow Transplantation

The technology of rare event analysis and sorting is applicable to a wide range of biomedical problems. In this paper we have discussed the application to prenatal diagnosis of genetic abnormalities through detection and isolation of fetal cells from maternal blood. Other application areas in progress in our lab include: a) detection, isolation and molecular characterization of rare metastatic breast cancer cells for expression of specific oncogenes, tumor suppressor and multi-drug resistance genes by reverse-transcriptase PCR (RT-PCR), b) detection and isolation of hematopoietic

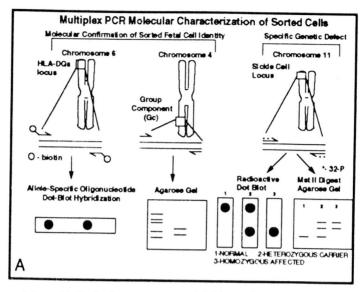

Figure 19. Multiplex PCR amplification and analysis
strategies.

stem cells in peripheral blood and bone marrow for
autologous transplantation following chemotherapy,
c) high-speed purging of metastatic breast cancer cells
from bone marrow for autologous bone marrow transplantation.

In detection of rare metastatic breast cancer, a
new project, we are faced with many of the problems of
finding fetal cells, namely that the markers used do not
have a very high specificity at a rare cell frequency
level of 10^{-4} to 10^{-6}. Thus we are using a strategy of
positive and negative selection markers (Fig. 21). It
is frequently wise to construct computer models of rare
cell and total cell data to determine what the approximate level of detection will be using particular cell
markers. Previously, we used the method to model rare
cell subpopulations on a single flow cytometric parameter (3). We have written software which now allows us
to do much more complex modeling with actual multiparameter listmode data from different cell populations

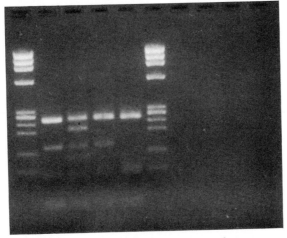

M   U   Hae III   Sty I   Hha I   M   Type 1F1S

HLA + Gc DIGESTION   ♂ DNA

Figure 20. Multiplex PCR gel for simultaneous detection
of HLA- DQα and Gc variant genotypes. Lanes 1 and 6 are
molecular weight markers. Lane 2 is uncut PCR product
containing the 242 bp HLA band and the 132 bp Gc band,
and lanes 3-5 are restriction digests of the product.
Lane 3:HaeIII, lane 4:StyI, and lane 5:HhaI. The HLA-
DQα product yields the band at 242 bp. The donor's Gc
type is 1F1S yielding a mixture of uncut (132 bp) and
cut (98+34 bp) fragments with HaeIII, and only the uncut
132 bp fragment with StyI. HhaI cuts all Gc types
except 1A1 and yields only cut (71 and 61 bp) fragments,
as expected. The additional band at ~200 bp is a digestion product of the PCR-amplifed DNA using HLA-DQα
primer #4 (Fig. 15B), which contains a HaeIII recognition site within the amplified region.

acquired in separate data files. The software allows
rapid determination of the detection limits of rare cell
subpopulations by performing a computer mixture at any
desired frequency of multiparameter data from two or
more listmode data files. The identity of each cell in
the data mixtures is tracked by use of additional classifier parameters (eg. cluster numbers from cluster
analysis). Such classifier parameters, not used by the

312

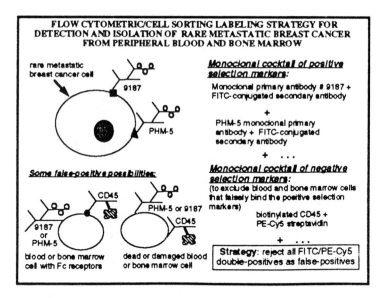

Figure 21. Schematic of breast cancer cell detection strategies. Sometimes multiple markers are needed to identify all of the rare cells of interest. In this case, it is more practical to reduce the numbers of colors of fluorescence markers by making monoclonal "cocktails" both for positive and negative selection markers.

multiparameter data analysis method being tested, are used to determine how accurate the data analysis method was in determining the correct identity of each cell. We are now performing such analyses of rare breast cancer cells in human bone marrow. In Fig. 22, we show such a modeling with a one percent mixture of MCF-7 human breast carcinoma cells labeled with a monoclonal antibody (#9187, a generous gift from Dr. Ping Law, Baxter Laboratories) generated from monoclonal antibody clone 260F9 (6) within a larger population of human bone marrow. This procedure is being used to determine the multiparameter labeling methods needed to successfully detect breast cancer cells at frequencies of less than 10^{-5} for minimal residual disease monitoring of breast cancer patients in remission after chemotherapy.

The quantitation of mRNA levels of specific oncoge-

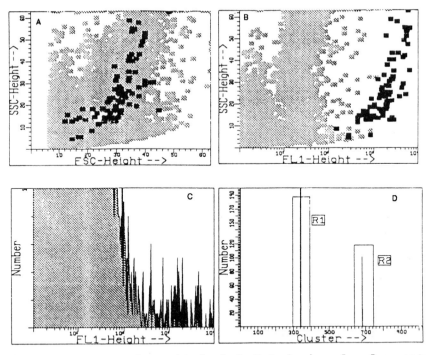

Figure 22. Sometimes it is helpful during development
of methods for specific applications to perform computer
mixtures of rare and non-rare cells to determine the
limits of detection of the rare cells. Data mixtures
shown in this figure (panels A-C with MCF-7 cells shown
in black) are computer mixtures of multiparameter listmode data, in this case mixtures of human breast cancer
(MCF-7) cells and human bone marrow cells labeled with
monoclonal antibody (#9187, Baxter Labs). To track the
identity of cells once the data are mixed, a classification parameter (R1 for human bone marrow cells and R2
for MCF-7 cells) has been added to each separate data
file before merging the data into a common data file
(panel D). Different data analysis methods for detecting the rare cells can then be quickly and easily tested
using the common data file.

nes, tumor suppressor genes, and multi-drug resistance

genes will be very difficult due to the need for purity,

yield, and the need for quantitative RT-PCR, itself

still a problematic technique especially for very low

numbers of cells. Sorted cells will also be examined

for chromosome rearrangements by chromosome FISH paint-

ing and confocal microscopy. The 3-D autostereoscopic
display is now being used in our laboratory to visualize
and analyze chromosome rearrangements in interphase
nuclei.

For the detection and isolation of hematopoietic
stem cells from peripheral blood and bone marrow, we are
using CD34 as a positive selection marker and a cocktail
of negative selection markers to exclude other cell
types as well as dead/damaged cells by dye exclusion, we
feel that use of a marker which should not stain any
cells except non-specifically is more likely to detect
cells staining non-specifically with antibodies. Thus,
all positive cells must also **NOT** stain for the non-
specific marker. At the level of ultra-rare cells many
PI-negative cells still bind antibodies non-specifically. This non-specific marker strategy should be more
sensitive than PI dye exclusion (Fig. 23).

The purging of rare metastatic breast cancer cells
from bone marrow to allow autologous bone marrow transplantation of a cleansed marrow involves high-speed
sorting with a strategy of "when in doubt throw it out"
since we can afford to throw away some normal cells but
we do not want to include any possible tumor cells.
Hence all positives and "not-sures" are removed and a
"flexible sorting" (31) anti-coincidence strategy is set
to exclude all sorting units containing any tumor or
"not-sure" cells which may be present among the several
cells present in a droplet under these high-speed conditions.

SUMMARY

Conventional flow cytometry and cell sorting can be
used, with some effort and difficulty to identify and to
isolate rare cell subpopulations. Appropriate modifications to the data acquisition and sorting system hardware and software make it practical to isolate rare and

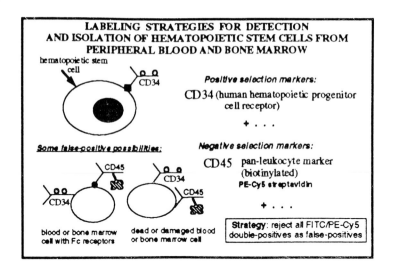

Figure 23. Schematic of hematopoietic stem cell detection strategies. If CD34 is used as the primary marker for hematopoietic stem cells (other markers can also be added), a large part of the problem becomes constructing an appropriate cocktail of negative selection markers to exclude all other cell types.

even ultra-rare cells. Secondary processing of rare sorted cells by image analysis/confocal microscopy, FISH and PCR for specific DNA sequences permit single-cell molecular characterizations of rare cell subpopulations.

ACKNOWLEDGEMENTS

The work presented in this manuscript was supported principally by Public Health Service Grant GM38645 with additional support by grants HD20601 and CA61531.

REFERENCES

1. Leary JF, Ellis SP, McLaughlin SR, et al: High-Resolution Separation of Rare Cell Types. In: Cell Separation Science and Technology, DS Kompala and P Todd (eds.), American Chemical Society Symposium Series No. 464, American Chemical Society, Washington, DC, pp. 26-40, 1991.
2. Jensen RH, Leary JF: Mutagenesis as Measured by Flow Cytometry and Cell Sorting. In: Flow Cytometry and Sorting (2nd edition), MR Melamed, T Lindmo T, and ML Mendelsohn (eds.), Wiley-Liss, New

York, pp. 553-562, 1990.

3. Cupp JE, Leary JF, Cernichiari E, et al: Rare-event analysis. Cytometry 5:138-144, 1984.

4. Ryan DH, Mitchell SJ, Hennessy LA, et al: Improved detection of rare CALLA-positive cells in peripheral blood using multiparameter flow cytometry. J. Immunol. Methods 74:115-128.

5. Frantz CN, Ryan DH, Cheung NV, et al: Sensitive detection of rare metastatic human neuroblastoma cells in bone marrow by two-color immunofluorescence and cell sorting. Progress in Clin. Biol. Res. 271:249-262, 1988.

6. Leslie DS, Johnston WW, Daly L, et al: Detection of breast carcinoma cells in human bone marrow using fluorescence activated cell sorting and conventional cytology. Amer. J. Clin. Path. 94:8-13, 1990.

7. Visser JW, De Vries P: Identification and purification of murine hematopoietic stem cells by flow cytometry. Methods in Cell Biology 33:451-468, 1990.

8. Cory JM, Ohlsson-Wilhelm BM, Brock EJ, et al: Detection of human immunodeficiency virus-infected lymphoid cells at low frequency by flow cytometry. J. Immunol. Methods 105:71-78, 1987.

9. McCoy JP, Jr, Chambers WH, Lakomy R, et al: Sorting minor subpopulations of cells: Use of fluorescence as the triggering signal. Cytometry 12:268-274, 1991.

10. Leary JF: Strategies for Rare Cell Detection and Isolation. In: Methods in Cell Biology: Flow Cytometry, Z Darzynkiewicz, JP Robinson, and HA Crissman (eds.), 42:331-358, 1994.

11. Ashcroft RG: Future clinical role for flow cytometry. (Review). Cytometry Supplement 3:85-88, 1988.

12. Kompala DS, Todd P: Cell Separation Science and Technoloqy. In: American Chemical Society Symposium Series No. 464, American Chemical Society, Washington, DC, 1991.

13. Peters D, Branscomb E, Dean P, et al: The LLNL high-speed sorter: Design features, operational characteristics, and biological utility. Cytometry 6:290-301, 1985.

14. Parson JD, Hiebert RD, Martin JC: Active analog pipeline delays for high signal rates in multistation flow cytometers. Cytometry 6:388-391, 1985.

15. Van den Engh G, Stokdijk W: Parallel processing data acquisition system for multilaser flow cytometry and cell sorting. Cytometry 10:282-293, 1989.

16. Herweijer H, Stokdijk W, Visser JW: High-speed photodamage cell selection using bromodeoxyuridine-Hoechst 33342 photosensitized cell killing. Cytometry 9:143-149, 1988.

17. Leary JF, Corio MA, McLaughlin SR: System for High-Speed Measurement and Sorting of Particles. U.S. Patent 5,204,884. International patents pending in Europe, Canada and Japan, 1993.

18. Blumenson L: SSIZE Computer Software (finite sampling theory), personal communication.

19. Gross D, Harris CM: Fundamentals of Queuing Theory, (2nd edition), John Wiley and Sons, New York, 1985.

20. van Rotterdam A, Keij J, Visser JW: Models for the electronic processing of flow cytometric data at high particle rates. Cytometry 13:149-154, 1992.

21. Jackson A, Warner NL: Manual Clinical Immunology, pp. 226-235, 1986.

22. van Dekken H, Arkesteijn GJ, Visser JW, Bauman JG: Flow cytometric quantification of human chromosome specific repetitive DNA sequences by single and bicolor fluorescent in situ hybridization to lymphocyte interphase nuclei. Cytometry 11:153-164, 1990.

23. Trask B, van den Engh G, Pinkel D, et al: Fluorescence in situ hybridization to interphase cell nuclei in suspension allows flow cytometric analysis of chromosome content and microscopic analysis of nuclear organization. Human Genetics 78:251-259, 1988.

24. Wade P: "COOKIE CUTTER" Method of Cell Selection. Meridian Instruments Application Note C-1, 1987.

25. Saiki RK, Bugawan TL, Horn GT, et al: Nature 324:163-166, 1986.

26. Ruano G, Brash DE, Kidd KK: PCR: The first few cycles. Perkin Elmer Cetus, Amplficatons 7:1-4, 1991.

27. Langauer C, Green ED, Cremer T: Genomics 13:826-828, 1992.

28. Telenius H, Carter NP, Bebb CE, et al: Genomics 13:718-725, 1992.

29. Zhang L, Cui X, Schmitt K, et al: Whole genome amplification from a single cell: Implications for genetic analysis. Proc. Natl. Acad. Sci. (USA) 89:5847-5851, 1992.

30. Gross H-J, Verwer B, Houck D, Recktenwald D: Detection of rare cells at a frequency of one per million by flow cytometry. Cytometry 14:519-526, 1993.

31. Corio MA, Leary JF: System for flexibly sorting particles. US Patent 5,199,576 (1993). International patents pending in Europe, Canada and Japan.

32. Buican TN: Automated Cell Separation Techniques Based on Optical Trapping. In: Cell Separation Science, DS Kompala and P Todd (eds.), American Chemical Society Symposium Series No. 464, American Chemical Society, Washington D.C., pp. 59-72, 1991.

33. Huber PJ: Annals of Statistics 13:435-525, 1985.

318

34. Gabriel CR: The biplot graphic display of matrices with applications to principal component analysis. Biometrica 58:453-467, 1971.
35. Nederlof PM, van der Flier S, Vrolijk J, et al: Fluorescence ratio measurements of double-labeled probes for multiple in situ hybridization by digital imaging microscopy. Cytometry 13:839-845, 1992.